Prescription
for the Planet

PRESCRIPTION FOR THE PLANET

"This is the most important book that has ever been written on sustainable development... You MUST read it! It is not A revolution, it is THE revolution, THE way to go!"

— *Bruno Comby, Ph.D., founder and President of EFN*
Environmentalists For Nuclear Energy

"If you're looking for an energy revolution, Blees has the boldness to offer both technology and vision."

— *Jim Hightower*

"Blees writes devilishly well. His book is a culmination of tremendous erudition compounded by no end of research. Whether our society can be turned around to follow his Pied Piper lead is open to question. But at least he's drawn a map."

— *T. J. King, Ph.D.*
Professor emeritus of English and Literature

"In a time desperate for solutions to the global environmental crisis, we need all the suggestions we can get. This analysis by Tom Blees therefore deserves serious attention as an informed and conscientious voice in the ongoing debate over what to do."

— *Howard Zinn. Professor, historian, playwright*
Author: A People's History of the United States

Please visit us at our web site:
http://www.prescriptionfortheplanet.com

ISBN: 1-4196-5582-5

ISBN-13: 9781419655821

Library of Congress Control Number: 2008905155

Visit www.booksurge.com to order additional copies.

Prescription for the Planet

The Painless Remedy for our
Energy & Environmental Crises

Tom Blees

This book is dedicated to my parents,
who taught me never to take no for an answer,
even though it sometimes made their parenting job
a lot more difficult, and who instilled in me
the confidence to look for answers
no matter how elusive they might seem.

Table of Contents

Introduction

No problem can be solved from the same consciousness that created it.
We must learn to see the world anew.
— Albert Einstein

I'D LIKE TO invite you to a revolution.

Don't worry, you're not going to get hurt. As a matter of fact, for the vast majority of the people involved it's going to feel really good. So good, in fact, that you'll wonder why this revolution isn't already underway.

The tensions that are leading up to it are visible all around us. Anyone who reads, listens to the radio or watches TV is barraged with dire warnings of environmental, political, and economic stresses almost mind-numbing in their complexity and portent. So how, one may ask, are the pressing problems of the day to be solved? Any revolution promising to deliver humanity from such disparate threats as global warming and resource wars will have to combine technical transformation on a par with the Industrial Revolution along with unprecedented political vision. As formidable as that sounds, it is entirely within the realm of possibility.

The political and technological solutions to a host of our

planet's most pressing problems are inextricably entwined. The common threads that unite many of them are energy and raw materials. Energy, in particular, is a nettlesome concern. The ways in which we source and use energy have profound effects on geopolitics, economics, and the environment. In the face of overwhelming evidence that business as usual is simply not going to work much longer, the search is clearly on for alternatives. Passionate advocates of various energy systems tout the virtues of their favorites in the media, most often with conveniently hazy statistics and projections. Whereas there seems to be a developing consensus that energy production and use are deadly serious issues, most of the purported solutions to energy problems continue to fall woefully short of the mark.

If this situation finds you frustrated or devoid of hope as you contemplate mankind's future, take heart. It will probably surprise you to know that there is a virtually inexhaustible source of energy that is safe, clean, and economical that will require no recourse to mining, drilling, or other extraction processes for literally hundreds of years. Far from being another pie in the sky, this technology was developed at one of America's national laboratories over more than a decade by a veritable army of PhDs. As the project reached its triumphant conclusion in the mid-90s, it was suddenly terminated and its facilities dismantled. The scientists who'd succeeded so spectacularly in their efforts were scattered, and word came down from the U.S. Department of Energy that the project was not to be publicized.

This is but one of a trio of little-known technologies that are capable — when coupled with prudent leadership — of solving a surprising array of seemingly intractable global problems. We'll start off with a brief discussion of the problems we seek to solve and then examine the pros and cons of the various purported solutions that have been suggested to remedy them. Beyond that we'll be breaking new ground, at least compared to what passes for conventional wisdom in today's public discourse.

When I speak of a revolution, I use the term advisedly. The course of action proposed herein will change the world every bit as profoundly as technological and political revolutions of the past. Unlike those social transformations, however, we are uniquely capable of planning this revolution in order to minimize the negative impacts of the changes it will bring about, and maximize its benefit to all of humanity.

By the time you've traversed these pages I believe you'll agree that we stand on the threshold of a new era in the evolution of human society. If we look back at the historical record, there's an unbroken and rather bleak consistency in the struggle for power over others, with wars of conquest evincing little substantive difference over the ages save for the methodology of slaughter. The thirst for riches and resources took a new turn once the Age of Exploration played itself out. From then on, there were no new lands to discover. Control over resources became a matter of wresting them away from someone else. Such a course was pursued with vigor during the era of colonization, but the end of World War II brought a new twist as warmaking technology — most obviously atomic weaponry — made wars of conquest a much dicier endeavor.

The proxy wars between the nuclear powers during the Cold War era can be seen as a relic of the old pattern, outmoded but alas, not yet abandoned. Even before the end of the Cold War it was clear that the struggle for control over ever more crucial supplies of resources would be played out on the stage of international economic relations. While we can still unfortunately see the brute force methods being used in the current war in Iraq, the relative stability of international borders portends a future where international trade and economic alliances decide who controls the world's raw materials. We can clearly see new tensions developing as China's burgeoning growth has made it a force to be reckoned with in the global struggle over energy supplies, even as those resources are revealing their limits as never before.

Do not despair. The struggle for control among an ever-increasing population for an ever-dwindling stockpile of needed materials is about to take on a new and encouraging dimension. We'll see in the pages to come not only how we can tap a limitless supply of environmentally benign energy already ours for the taking, but how to effortlessly recycle nearly everything that provides us with the comforts of life we now enjoy.

Ever since our planet's physical limitations were recognized, the relationship between nations was based on the concept known today as zero-sum. As the most advanced industrialized nations consume an inordinately large share of the world's resources, the threat that the rest of the countries of the world will eventually demand their fair share looms on the horizon. A zero-sum world can be likened to sharing a pie: if you take a bigger slice, somebody else is going to have to take a smaller one. The lack of enthusiasm for helping to lift the poorest nations out of their misery can be traced to the nagging fear that enlarging their piece of the pie will inevitably diminish what's left for the rest of us.

This resigned acceptance of the zero-sum paradigm is still in evidence virtually everywhere we look today. Neither the public nor the political class has yet recognized that this way of thinking is already obsolete. Few are yet aware that the pie hasn't just gotten bigger. We're looking through the window of the pie shop, just waiting for the world's leaders to show up with the keys. Inside there's more than enough for everybody.

Mankind is poised on the brink of a new age of plenty. The wealthiest nations need not fear that elevating the poor of the world will diminish their own standard of living. On the contrary, improving the condition of the poorest among us will improve everyone's situation if only because it will greatly diminish the inevitable tensions resulting from gross inequality. Access to abundant and affordable energy supplies will no longer be the prerogative of the fortunate, but will finally be recognized as a basic human right.

This invitation to revolution is not a call to arms. It is a call to action. We have the means to radically transform human society for the better while solving some of the most formidable problems humanity has ever faced. What we need is the vision and the will to implement this global revolution, one whose effects will impact the lives of all the world's people in unexpected and gratifying ways. Let us begin...

CHAPTER ONE

A World of Hurt

*There are good people... who hold this at arm's length
because if they acknowledge it and recognize it then the
moral imperative to make big changes is inescapable.*

— Al Gore, *An Inconvenient Truth*

A S THE TWENTIETH century drew to a close there was much talk about the challenges facing mankind as we began the new millennium. Now just eight years past that milestone, many of those issues have taken on a startling urgency. While the end of the Cold War brought relief at the diminished threat of nuclear annihilation, new threats until recently only dimly perceived have taken its place. The danger of nuclear warfare between two great powers has been supplanted by the specter of nuclear proliferation. And the dilemma of human-caused global warming is regarded by virtually every nation as a grim reality and one of the most daunting challenges humankind has ever faced.

The greatest difficulties we face today are nearly all of our own making. We have burdened the planet not only with our sheer numbers but with the ability to profoundly influence our environment with advanced technology. Our booming population exacerbates the situation in both industrialized and undeveloped countries. In the former the deleterious effects of development pollute both air and water, sometimes to unprecedented degrees. In undeveloped nations, the sheer demand for living space and simple fuel leads to extensive deforestation and both indoor and outdoor pollution. It has gotten to the point where we have the very real possibility of despoiling our planet so severely that human life itself, if not imperiled in its very existence, seems to be approaching the point of serious social disruption.

For most of the twentieth century, there was a widespread belief in science's ability to unravel and solve our world's technological and environmental problems. The irony is that scientific advancements were creating whole new problems that had never existed before, leading many to question whether science has been a panacea or a Pandora's box. Today the number of people who blithely assume that scientists will be able to sort it all out in time seems to be inexorably diminishing. Indeed, a backlash of anti-science forces have found, at the time of this writing, a sympathetic administration in Washington which at least pretends — for the sake of their votes — to share their antipathy to what many of them see as the scientific priesthood.

Like an environmentalist driving his SUV to a global warming conference, America's neo-Luddites avail themselves of the comforts of their technological cocoon even as they attempt to eat it away from the inside. Such inconsistency and irrationality would hardly be worth confronting except for the political results that are postponing the recognition and solution of serious environmental problems. An improbable alliance of anti-science zealots on the one hand and environmentally callous corporations on the other has thwarted progress on a host of issues which frankly can't afford to be ignored any longer.

Dozens of books and countless articles have been written about the grave challenges briefly described below. My intention in this book is not to expound on and lament the problems that bedevil us but to offer realistic solutions. But first we must identify the targets. This first chapter will briefly present the issues that cry out for solutions. Every one of them, as incurable as they may seem, will be addressed in the chapters to come with a comprehensive plan to remedy them in the near future without resorting to technological leaps of faith.

Be forewarned: Once you finish this book and realize that there are actually completely feasible near-term solutions to these problems, it may drive you nuts listening to the pundits and "experts" on radio and TV pontificating on these issues and how they propose to address them. You'll read an article on global warming or alternative energy systems or clean coal or biofuels and it will sound remarkably akin to that old story about the blind men and the elephant. Early readers of this manuscript have told me they're tearing their hair out at the barrage of gloom and doom and solemn pronouncements, now that they've discovered the planetary prescription. Don't say I didn't warn you.

Global Warming: The elephant in the room

Climate change seems an amorphous and intangible concern to most people. But the Inuit people of Baffin Island, which sits atop Canada just west of Greenland, have gone beyond debating the reality of global warming. While politicians in their comfortable offices dicker over the science, the way of life of the Inuit who've lived on Baffin Island since time immemorial is being destroyed by unprecedented warming of their environment. Where once they hunted on the ice for ten months a year, now their hunting season has been reduced to about half that time. The evidence of a drastically altered climate is all around them, and it is altering their culture to a profound degree.[1]

Further south, however, the evidence is somewhat less immediate and thus the implications of global warming have taken longer to recognize. Nevertheless, concern over the possible threat of human-caused climate change led to the establishment of the Intergovernmental Panel on Climate Change (IPCC) in 1988. A collaboration between the United Nations Environment Program and the World Meteorological Organization, the IPCC was created to assess the risk of human-induced climate change based on the best scientific and technical information available.

Nearly two decades after its creation, the IPCC's pronouncements find themselves the focus of world attention. In February of 2007 the panel issued the first installment of their report on climate change, the culmination of the last six years' work of some 2,500 scientists around the world. Their "Summary for Policymakers" reported the verdict that it is "very likely" that human activities (in particular the burning of fossil fuels) account for most of the warming in the past fifty years. "Very likely" translates as at least a 90% degree of certainty.[2]

Nevertheless there were dissenters. Of the 113 countries participating in the IPCC conference in Paris that issued the report, there were unsuccessful attempts to water it down by Saudi Arabia (the world's largest oil exporter) and China, which has recently overtaken the USA as the world's worst offender in emissions of greenhouse gases (GHGs). The difficulties of crafting a consensus among so many nations resulted in an inevitable softening of the report's nonetheless compelling conclusions. Much of what is discussed freely and credibly among the scientific com-

---

[1] Will Steger, *Global Warming 101.Com* (Will Steger Foundation, 2006 [cited 2007]); available from http://www.globalwarming101.com/content/view/545/88889028/.

[2] Working Group 1 of the IPCC, "Climate Change 2007: The Physical Science Basis," (Geneva, Switzerland: Intergovernmental Panel on Climate Change (IPCC), 2007).

munity never made it into the final draft, despite considerable sound science underpinning substantially scarier observations:[3]

- Emerging evidence of potential feedback effects and "tipping points" that could rapidly accelerate global climate change;
- Growing proof that the Greenland ice sheet is melting at an increasing rate and could collapse entirely;
- Findings that temperatures in Antarctica are rising "faster than almost anywhere on the planet" and that the ice there is also in increasing danger of breaking up;
- Measurements of the Atlantic Gulf Stream, which plays a major role in the climate of Western Europe, revealing a 30% slowing between 1957 and 2004;
- The potential effects of accelerating release of greenhouse gas in the Arctic from thawing soil, permafrost and seabed deposits;
- The potential for dramatic and extreme rises in sea level should ice sheets continue to break up.

Undeterred by the consensus of some 2,500 of the world's top scientists, the incorrigible ExxonMobil quickly came up with a bounty of $10,000 to any scientist willing to poke holes in the report, albeit under the nearly transparent cover of a company-funded neocon think tank.[4] It would be futile to expect unanimous agreement about the realities and dangers of global warming among politicians. Yet a majority of those with the most comprehensive training in the subjects involved (oceanographers, climatologists, paleobotanists, etc.) appear to agree that mankind is affecting the cli-

---

[3] David L. Brown, *What the IPCC Report Didn't Tell Us* (2007 [cited 2007]); available from http://starphoenixbase.com/?p=353.

[4] Ian Sample, "Scientists Offered Cash to Dispute Climate Study," *The Guardian,* Feb 2, 2007.

mate in serious and potentially irreversible ways. They differ mainly in degree (no pun intended) when it comes to just what point we find ourselves at now and what the future holds, but you'd be hard pressed to find any of them who'd suggest that solutions to the problem are something we can afford to put off till tomorrow.

The belief that anthropogenic (human-caused) emissions of global warming gases are causing or exacerbating global warming is not absolutely universal among scientists. The subject is extremely complex, and some perfectly sincere scientists, not just paid shills of fossil fuel corporations, look at the evidence they have in hand and come to different conclusions. That the earth is experiencing a warming trend is hardly refutable, and the vast majority of scientists would find no quarrel with the evidence. Just how much of that warming trend is due to anthropogenic emissions, however, evokes less unanimity, though dissenters from that view are in a distinct minority. Nevertheless, this is a classic example of the scientific method at work. Evidence continues to accumulate, and by now it's gotten to the point that the leaders of many countries are sounding the alarm.

This book will frequently refer to the urgency of climate change as a reason to take decisive action to revolutionize the world's energy systems. While this is consistent with the views of the majority of scientists, some may beg to differ. Global warming is not the only reason, however, for the energy revolution that will be explained and encouraged in these pages. There are, in fact, a host of compelling reasons to initiate and carry out the program recommended here. If anthropogenic emissions end up being inconsequential (and it wouldn't be the first time that a large number of people, scientists included, may have had a shared misconception), we'll still have proceeded along a path that leaves us in a much better condition than if we had not, with substantial improvement on a host of other issues.

If we took the proposed path and the people on the planet suddenly had a change of heart en masse and decided to limit the size of their families, AND anthropogenic emissions turned out to be inconsequential, AND if the current warming we're experiencing halted and reversed itself, then would this course of action have been for naught? Not at all. As we shall see in the pages to come, we still would have spent less than if we'd taken a business-as-usual approach, we'd still have remedied the deadly problem of air pollution, and we'd still have more than enough energy resources for everyone on the planet. Ultimately the rationale for pursuing this course stands firmly on its own merits. If the reader looks with skepticism at forthcoming references to the urgency of global warming, please bear in mind that it is but one of many compelling reasons to pursue this energy revolution.

In the event that anthropogenic emissions are indeed as consequential to our climate dilemma as most scientists believe, then taking prompt action will certainly turn out to be the wisest course. In the unlikely case that mankind is not at least partly responsible, should the current warming trend persist for much longer there will be ample reason to pursue an energy strategy like the one that will be proposed herein. For the human population of the planet is growing toward a predicted peak of about ten billion,[5] even as glaciers that supply water to hundreds of millions of people are rapidly retreating. Not only will we have to supply billions more with fresh water (which will require a lot of energy), but there's a very high likelihood that hundreds of millions will soon find themselves displaced because of vanishing water supplies.

The accelerated melting of glaciers all over the globe is probably the most visible sign of global warming. To cite just one example, up to 64% of China's glaciers are projected to dis-

---

[5] "Total Midyear Population for the World: 1950-2050," (U.S. Census Bureau, 2007).

appear by 2050, putting at risk up to a quarter of the country's population who are dependent on the water released from those glaciers.[6] That's about the same number of people as inhabit the entire United States.

A look at almost any area of the world today where there are glaciers and/or ice caps reveals a rate of melting unprecedented in history.[7] From China to the Arctic, from the Andes to the Himalayas, the rate of glacial retreat is so dramatic that entire regions are in danger of losing their glaciers altogether. The water supplies which depend on those glaciers as their source will disappear, in many cases causing catastrophic disruptions among the countless millions of people who depend on them. Peru and Bolivia, which together account for more than 90% of the world's tropical glaciers, have lost about a third of the surface area of their glaciers between the 1970s and 2006. With three-quarters of Peru's population living on the arid west side of the Andes where less than 2% of that nation's water resources are found, the consequences of diminishing runoff are already starting to be felt.[8]

The economic costs of global warming are already visible, but the projections as global warming continues are truly staggering. Insurance industry estimates predict that climate-change related damages might cost $150 billion annually within a decade.[9] If the connection between the increased frequency and severity of hurricanes in recent years is partly a result of global warming, as many climatologists claim, then

---

[6] Renato Redentor Constantino, "With Nature There Are No Special Effects," in *TomDispatch.com* (June 3, 2004).

[7] Robert S. Boyd, "Glaciers Melting Worldwide, Study Finds," *Contra Costa Times,* Aug 21, 2002.

[8] James Painter, "Peru's Alarming Water Truth," in *BBC News International Edition* (Mar 12, 2007).

[9] Constantino, "With Nature There Are No Special Effects."

the tens of billions of dollars worth of damage from the hurricane strikes of 2005 alone in the United States is already pushing that estimate far closer than that decade estimate would suggest.

There are disturbing signs that we may have already reached a tipping point beyond which serious disruptions to the global climate are irreversible. Melting of previously stable permafrost is but one of the warning signs.

> Western Siberia is undergoing an unprecedented thaw that could dramatically increase the rate of global warming. Researchers recently returned from the region found that an area of permafrost spanning one million square kilometers—the size of France and Germany combined—is melting for the first time since it formed 11,000 years ago at the end of the last ice age. British and Russian scientists report that the melting permafrost is releasing hundreds of millions of tons of methane, which is 20 times more potent than the carbon dioxide currently driving the worldwide warming crisis.[10]

Sergei Kirpotin, a botanist at Tomsk State University, Russia, describes an "ecological landslide that is probably irreversible and is undoubtedly connected to climatic warming." He says that the entire western Siberian sub-Arctic region has begun to melt, and this "has all happened in the last three or four years."[11]

To anyone who pays attention to scientific periodicals or even general news sources, the number of studies attesting to

---

[10] Ian Sample, "Warming Hits 'Tipping Point'," *The Guardian,* Aug 11, 2005.
[11] Fred Pearce, "Climate Warning as Siberia Melts," *New Scientist,* Aug 11, 2005.

the reality and urgency of global warming is overwhelming. Reports by scientists from a variety of disciplines continue to pour in from around the globe. One day it's a story of an Antarctic ice sheet the size of Texas starting to disintegrate. Then a story that the glacier on Mount Kilimanjaro that started growing almost 12,000 years ago will probably be gone within a decade or two. Polar bears are dying because they can't navigate the ever-widening gaps in the ice floes as the Arctic ice melts away. The Atlantic thermohaline circulation, which is responsible for the currents that warm northern Europe, may even be slowing down.[12] These are hardly subjective assessments. Cold hard data is pouring in from around the world, bringing incontrovertible evidence that we've created a problem the likes of which mankind has never before had to face.

The European Project for Ice Coring in Antarctica (EPICA) team has spent years drilling the ice core in Antarctica's Ice Dome Concordia. They had previously analyzed its record of global temperatures, but have just completed the detailed analysis of the trapped air. The bubbles record how the planet's atmosphere changed *over six ice ages and the warmer periods in between* [my italics]. But during all that time, the atmosphere has never had anywhere near the levels of greenhouse gases seen today. Today's level of 380 parts per million of carbon dioxide is 27% above its previous peaks of about 300 ppm, according to the team led by Thomas Stocker of the University of Bern in Switzerland.[13]

---

[12] Michael Mann Gavin Schmidt, "Decrease in Atlantic Circulation?," in *Real Climate* (Nov 30, 2005).

[13] David L. Chandler, "Record Ice Core Reveals Earth's Ancient Atmosphere," *New Scientist,* Nov 24, 2005.

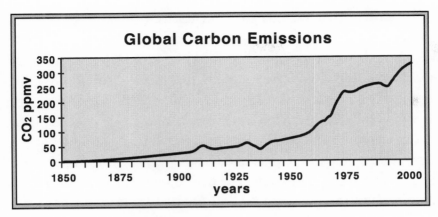

A thoroughly modern problem [14]

Global warming alone is reason enough to warrant a radical and comprehensive overhaul of energy production and use throughout the world. Whether the potentially disastrous effects of climate ochange can be reversed or at least halted somewhere short of disaster is an open question. At this point we can only do as much as possible to halt the human practices that are contributing to the ever-deteriorating climate situation.

One study after another, whether by international groups of esteemed scientists or studies done by the scientists of individual nations, points to the same conclusion. Despite the protests of intransigent politicians in the United States and their apologists, along with their often uninformed believers among the general populace, global warming is not really a question of if but rather of how seriously and how quickly it will manifest.

Those who choose to believe a small minority of the scientific community when their views contradict the evidence and studies of the vast majority have no business formulating

---

[14] (Delft Technical University) Lenntech, "The Global Warming and the Greenhouse Effect," (2007).

public policy that will impact the entire human race. Yet several powerful politicians—to our global shame, mostly in the United States—still pretend that global warming is an environmentalist conspiracy. Senator James Inhofe (R-Oklahoma), who ironically chaired the Senate Environment and Public Works Committee until mercifully being ejected from that position by the 2006 elections, has called global warming "the greatest hoax ever perpetrated on the American people."

Many have castigated the U.S. government for dismissing the Kyoto Accords on Global Warming, resisting for years even the most rudimentary admission of the reality of climate change, much less the causes. To be sure, the signal this sends to the rest of the world is deplorable, yet the Kyoto Accords were only a very feeble first step that, even if embraced, would hardly turn the tide. We must go far beyond the reach of Kyoto to address global warming, and we have to do it faster than that agreement would have demanded. Alas, many of our politicians seem to be headed in the opposite direction.

While many hoped for real progress at the 2006 U.N. climate summit in Nairobi, Kenya, it ended instead with disappointment and failure. The intransigence of the United States and China, the two most egregious producers of greenhouse gases, doomed the conference despite the high hopes of its other participants. It's now generally recognized among the world community that the Bush administration is determined to shirk its responsibility. "Everyone is waiting for the [U.S.]," said Paal Prestrud, head of Oslo's Center for International Climate and Environmental Research. "I think the whole process will be on ice until 2009 [when Bush will be replaced]."[15] It is not known whether Mr. Prestrud appreciated the cold irony of his choice of words.

---

[15] The End Is Sigh (Grist Environmental News & Commentary, Nov 20, 2006 [cited); available from http://www.grist.org/news/daily/2006/11/20/.

Shortly after this book goes to print, George Bush will be leaving the White House. Those who have decried U.S. foot-dragging on global warming will find herein a comprehensive plan to halt anthropogenic emissions of greenhouse gases much faster and more thoroughly than any international plan to date. But that is only one of the issues we will address. A surprising array of seemingly intractable problems facing us today can actually be solved with a small suite of bold actions that fit together like the workings of a classic timepiece. Arresting global warming would simply be icing on the cake.

Nuclear Proliferation

Americans who grew up in the Fifties and Sixties developed a particular knack for relegating worries about nuclear weapons to our mental closet of horrors. Never before had a whole generation of children been forced to undergo nuclear attack drills, rushing out of our classrooms to hunker down in the hall, sit on the floor and, as the macabre joke of the time described it, "put your head between your legs and kiss your ass goodbye." One wonders how much the threat of imminent annihilation contributed to the culture of hedonism which came to prevail in the hippie era of the Sixties and Seventies. It seems incongruous to think that "Eat, drink, and be merry, for tomorrow we die" originated in the Old Testament, since it could well have been the motto of young Americans who came of age in those perilous years.

Even though most of that generation is grown now with children and even grandchildren of their own, nuclear proliferation is still one of those awesome threats that most people refuse to contemplate. The end of the Cold War seemed to bring a welcome relief from such concerns, yet the nuclear bogeyman refuses to go away. Not only is the "nuclear club" growing, but terrorism has worked its way to the forefront of international concerns, along with the very real possibility that eventually a terror attack will include the horrific prospect of a city suddenly vanishing in a nuclear explosion.

Some needed attention has focused on the lax control over nuclear weapons stockpiles as a result of the breakup of the Soviet Union. But as North Korea elbowed its way into the nuclear club in 2006 a more insidious threat reared its head. For the Koreans had created their first nuclear weapons not from stolen weapons-grade material but, following India's example, by operating a small reactor in such a way as to produce weapons-grade plutonium. There are probably several hundred tons of weapons-grade plutonium in existence, most of it (one hopes all of it) in the weapons programs of the nuclear powers. However, the technology to extract plutonium from spent reactor fuel is available to at least thirty countries, and any reactor can be adapted (at a sacrifice in operating efficiency) to production of weapons-grade plutonium.[16]

The "waste" (used fuel) produced during the course of a year by a normally operating nuclear power plant contains about 200 kilograms of low-quality, "reactor-grade" plutonium. Since that material can theoretically be used to make a nuclear explosion, it should certainly be safeguarded. Yet the emphasis that has been placed on weapons proliferation from spent power plant fuel is exaggerated, for its isotopic composition makes it unsuitable for weapons. There are far easier ways of producing weapons-grade material.[17]

As this is being written, America is rattling its sabers loudly over the prospect of war with Iran. While there is a multitude of possible reasons why — not the least of which is oil — Iran's development of uranium enrichment technology is most often cited as a *casus belli* by the Bush administration. Even as stalled talks to convince North Korea to abandon its nuclear weapons

---

[16] Bernard L. Cohen, "The Nuclear Energy Option," ed. University of Pittsburgh (Plenum Press, 1990).

[17] Ibid.

program have finally begun to bear fruit, the Iranians threaten to unleash the nuclear genie. It's like we're playing nuclear Whack-a-Mole.

The threat of nuclear proliferation has been with us since World War II, but the spread of modern technology has made it all the more urgent. Like all the problems that will be discussed here, this too is within our power to solve. The question is whether the world's leaders are willing to make the unprecedented decisions necessary to get the situation under control. As we'll see in the chapters to come, the international structures needed to eliminate the threat of nuclear proliferation — and global warming, and air pollution, and nuclear waste — are destined to collide with an international corporatism that has spread its tentacles into every corner of the globe.

What we're faced with at the dawn of the twenty-first century is a struggle for our very survival, but the struggle is not against some hostile outside force. It is against our own institutions, our own inertia, a dearth of imagination, a fear of change, and a selfish timidity on the part of our leaders.

A refusal to confront problems head-on has rarely promised such dire consequences as today. Fossil fuels are being burned at an accelerating pace, and unless revolutionary changes are made we will all be punished for our indecisiveness. The spread of nuclear weapons likewise must be recognized as the grave threat that it is. If one of our cities suddenly disappeared in an unexplained nuclear explosion, proliferation would immediately be front and center and the hue and cry for action would be deafening. We have to muster the good sense and the boldness to deal with this threat before such a horrific event occurs. Without radical changes to the way nuclear materials are handled, it will only be a matter of time. The longer we wait, the harder and more dangerous it will be to prevent such a catastrophe. It's time we recognize its inevitability and do everything in our power to get the situation under control.

Humans have a long and inglorious history of locking the barn door after the horse is gone. How many times have you heard of some local people insisting on the installation of a traffic light at a dangerous intersection, only to have the authorities drag their feet until someone is killed in an accident that could have been so easily prevented? The new traffic light that immediately appears might as well be a flashing tombstone. The same sort of oblivious inaction has gripped the world at large when it comes to dire warnings of nuclear weapons proliferation. No, the solution is not as easy as installing a traffic light, it will require bold leadership and a willingness to break free of old ways of thinking. But if we fail to act, it won't be a single tombstone that we'll be planting.

Air Pollution

The center of Mexico City is the Zocalo, with the National Cathedral on one side and the National Palace on an adjacent side. It's a one square block open area, a big park for residents and visitors alike to stroll and mingle. In my repeated visits to Mexico City over the years I can remember many days when I would enter the Zocalo from the street opposite the side where the palace sits. Looking across at the great edifice that occupies the entire side of the square, I could see only its outline. The massive doors and windows facing the park — a mere block away — were completely indistinguishable because of the thick smog.

Take the most complacent anti-environmentalist you can find and plunk him down in the middle of Mexico City (or any of a huge number of cities around the world) on almost any day of the year. Even if he's blind he'll still be struck by the pollution assailing his nostrils and lungs. Whatever a person might believe or disbelieve about global warming and the effect of human activity on climate change, only a raving lunatic would deny that air pollution in our major cities is a serious problem.

Like many of the environmental dilemmas facing us today, air pollution is a product of both our technology (and paradoxically, often also a lack of technology) and our sheer numbers. The concentration of humanity in urban centers is an inescapable fact of life, and it is increasing every year. It would be wonderful but hopelessly naive to think that people around the world will recognize the limitations of our biosphere in the very near future and stop their excessive procreation. We can count on adding at least a few billion more bodies to our already overburdened planet before the tide of humanity has a realistic chance of subsiding. Barring widespread nuclear war, unprecedented famine, or a deadly pandemic — either natural or man-made — we're stuck with the task of solving grave pollution problems despite the burgeoning population of our planet.

The causes of our deteriorating air quality are many and varied. With seemingly no sense of irony, people decry pollution caused by automobiles and lament the death of the "environmentally friendly" electric car. Yet the electricity for charging it more likely than not would originate at a coal-fired power plant, belching not just global warming gases like carbon dioxide into the air, but a host of other nasty substances as well. Sulfur dioxide emissions from coal burning have decimated large expanses of forests and made some lakes so acidic that all their fish died off. Mercury and lead emissions wafting from the smokestacks of coal-fired power plants have long been a concern because of their potential impact on child development.[18]

The urgency of finding a quick solution to air pollution worldwide is graphically illustrated in the case of China. As formerly "Third World" China becomes an industrial powerhouse and its people acquire the level of wealth necessary for modern conveniences, China's energy appetite is soaring. Even

---

[18] Cat Lazaroff, "Coal Burning Power Plants Spewing Mercury," in *Environment New Service* (Nov 18, 1999).

now, a third of China is bathed in acid rain on a regular basis due to coal-fired power plants, with over half its cities affected. Yet in order to meet their expected needs for electricity, China has dozens of coal-burning power plants on the drawing board to be built over the next few decades. If all these are brought on line as planned, the amount of pollution and global warming gases produced during their service lives will rival the entire world's current output. And India, whose population is set to outstrip China's during that time period, is likewise developing a ravenous energy appetite.

Even though coal burning tops the list, the most visible villain in the air pollution drama is the automobile. Despite strict emission control regulations and state-of-the-art systems on modern cars, the sheer number of vehicles on the road in many urban areas results in dangerous amounts of air pollution, especially when natural weather patterns conspire to create inversions. Climatic inversions occur when a warm body of air moves in over a cooler, denser body of air closer to the ground. The result is almost as if a lid were put over the area, trapping pollution in the cooler ground layer, often for days at a time. It's even worse in countries that lack the legal or financial means to enact and enforce emissions controls.

My experience on a recent trip to India can serve as one small example of the problem. I'd hired a car in Agra, home of the Taj Mahal, to take my son and me to the Himalayas. Agra has enacted more stringent auto emissions standards than almost anywhere else in India because of the very real possibility of acid rain slowly dissolving the stone of the Taj Mahal. Midway through our trip our driver's diesel car (very common in India) developed a problem with its catalytic converter, an integral part of a car's pollution control system. How did the mechanic deal with the problem? He removed the catalytic converter, smashed and emptied its innards, and placed the empty shell of it back on the car. Could one

realistically expect that this expensive part would be replaced any time in the near future? Doubtful at best. Mul-
tiply that vignette — or worse — repeatedly in developing
countries around the world.

   Things have gotten so bad in south Asia that we've seen
the development of what has been termed The Asian Brown
Cloud. (When representatives of countries under the cloud
complained that the term unfairly stigmatized them, the P.C.
police renamed it the Atmospheric Brown Cloud, apparently
so they could keep the catchy ABC acronym. In the interest of
clarity and at the risk of seeming politically incorrect, I will
refer to it hereafter by its original moniker, since it simply
indicates the cloud's location.) A team of over 250 scientists
from the U.S., Europe, and India completed intensive field
observations in south Asia in 1999 and were stunned at what
they found.

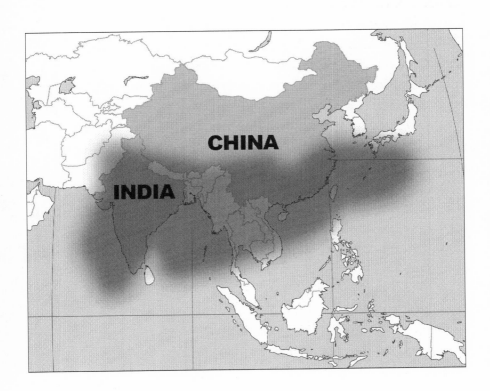

When the researchers first began noticing this smoggy haze, they thought it might be confined to major cities. As it turns out, it's an enormous blanket covering much of the area around the northern Indian Ocean. This part of the world is home to nearly 3 billion people, or about half the world's population, and it's industrializing rapidly. And because these countries can't afford state-of-the-art, energy-efficient technology, most of the new industries there are using old-fashioned, highly polluting engines and fuels.[19]

For at least a few months every year this brown haze hangs over most of south Asia, from Afghanistan in the west to the south of Japan. Not only is it hazardous to the health of the people living beneath the two-mile thick layer, but the haze scatters sunlight and reduces evaporation from the ocean, leading to less rainfall in an area of the world that can hardly afford it.

> "It's made of a variety of nasty substances, including fly-ash, sulfuric acid, particles from the burning of diesel and other fuels... it is extremely unhealthy and is also having quite important impacts on weather systems."
> — *Nick Nuttall, UN Environment Program* [20]

What may seem surprising to inhabitants of modern industrialized nations is the fact that a large part of the brown cloud comes from millions of people burning wood or dung in their homes for cooking.[21] Clearly the problem of air

---

[19] Bob Hirshon, "Asian Brown Cloud," in *Science Netlinks* (AAAS) (Jan 12, 2003).

[20] Radio Netherlands, "Brown Pall over Asia," (Aug 12, 2002).

[21] Ibid.

pollution, while differing in its sources and composition depending on the country and the season, is a deadly serious one for most of earth's inhabitants. A 1997 joint study of the World Health Organization (WHO), the World Resources Institute (WRI) and the US Environmental Protection Agency (EPA) estimated that annually nearly 700,000 deaths worldwide are related to air pollution and that this number may escalate to 8 million deaths by 2020.[22]

Nuclear Waste

The term nuclear waste is actually somewhat of an ironic misnomer. Most people assume that the reason it's considered waste is because all its usable material has been removed. In reality, not even 1% of the uranium ore's potential energy is used in a conventional light water reactor (LWR) or heavy water reactor (HWR), variations of which comprise nearly all of the reactors in use today.[23] If this seems like an incredible waste, then you can see the double entendre of the term quite clearly. The problem lies not only in the fact that we're throwing away so much fuel, but that what we're discarding creates an environmental legacy that will be hazardous to our progeny virtually forever.

In addition to the nuclear waste from reactors, the countries of the world that possess nuclear weapons have amassed a large quantity of weapons-grade material that has been recycled out of old warheads and is in need of disposal. So far about 260 tons of it have been produced, mostly by the nations of the "nuclear club," with more being produced all the

---

[22] UNEP (United Nations Environment Program) Assessment Report, 2002. *The Asian Brown Cloud: Climate and Other Environmental Impacts*

[23] George S. Stanford, "Integral Fast Reactors: Source of Safe, Abundant, Non-Polluting Power," in *National Policy Analysis* (Dec 2001).

time.[24] Some has been reprocessed into so-called MOX fuel to burn in nuclear reactors, but between that and the much greater quantity in spent fuel from nuclear plants we face a dilemma hitherto unknown to mankind.

The United States is the reluctant owner of much of the world's nuclear waste. After World War II the U.S. started its Atoms For Peace program, exporting nuclear technology for peaceful purposes (and undoubtedly for the benefit of the U.S. nuclear industry). Not wanting to have all that nuclear material scattered around the globe, however, the Americans stipulated that the 41 countries that participated in the program would have to ship their waste back to the USA.[25] It sounds more than a little naïve (and uninformed) when cries of alarm are heard about moving nuclear material around within the country, since we've been shipping the stuff all around the world with relative impunity for half a century. At this point, between its own production and the leftovers from its atomic client states, the U.S. is trying to come to grips with about 50,000 tons of used nuclear fuel.[26] Though the Atoms For Peace program was abandoned long ago, the policy of using the United States as a dumping ground for the countries involved continues to this day.

Despite this grim situation, an ever-increasing number of people are advocating a wholesale embrace of nuclear power regardless of the waste it generates, out of sheer desperation to stop the progress of global warming. Even some longtime icons of the environmental movement are now speaking up as advocates of nuclear power, and of course the nuclear industry

---

[24] William M. Arkin Robert S. Norris, "World Plutonium Inventories–1999," *Bulletin of the Atomic Scientists* Sept-Oct 1999.

[25] "Spent Nuclear Fuel Returned to the United States from Germany," ed. U.S. Dept of Energy (National Nuclear Security Administration, Sept 2004).

[26] Public_Citizen, *New Nuclear Power Plants = More Nuclear Waste* (Aug 2003 [cited]; available from **http://tinyurl.com/5lps7a**.

is doing its best to be there with designs for a new generation of reactors. Nevertheless, disposing of the prodigious
amounts of nuclear waste that we've already produced is a
tall order that's generated immense controversy.

Even the newly converted are largely unaware that nuclear
waste need not be a problem any longer. The grudging acceptance of the hazards of long-lived nuclear waste in exchange for
addressing the global warming crisis is a Faustian bargain that
need not be transacted. We'll see in the pages to come how we
can avoid leaving a legacy of nuclear waste to future generations
by turning a worrisome liability into a valuable asset.

Oil Shocks

At the time of this writing (and hopefully not at the time of
your reading), the United States is deeply immersed in war
in Iraq. Despite the obvious involvement of oil as a major
factor in this war, there are some who would argue that the
oil involved — generally reported as the second largest oil
reserves of any nation — was not a causative factor in America's aggression. Be that as it may, it is clear that numerous
wars have been fought over fossil fuel resources, either as
the main reason (as in the first Gulf War under Bush Sr.)
or, more frequently, as an undeniable element in either the
progress or the triggering of hostilities, such as the Japanese
invasion of the Dutch East Indies early in WWII.

Even when not contributing to all-out warfare, the un-
equal distribution of natural resources contributes dispro-
portionately to international tensions, and few such resources
create more tension than energy supplies. Just witness the political stresses between the USA and the government of Ven-
ezuela, one of the Americans' major suppliers of oil. President Hugo Chavez claimed that the failed coup against him
in 2002 transpired with the cooperation, if not the instiga-

tion, of the United States.[27] Protestations to the contrary by
the Bush administration were rendered somewhat suspect by
the alacrity with which the USA recognized the coup's leaders, who held power for only two days before Chavez was
reinstated.

Even as futurists are predicting wars that will be fought
over water in the not too distant future, we have already been
embroiled in fossil fuel wars for decades. The geopolitical instability caused by a desire for control of such resources is arguably
one of the greatest impediments to peace in the world. As in-
dustrialization and prosperity spread to previously undeveloped
nations, the competition for energy grows ever more serious.

> We are on the cusp of a new kind of war — between
> those who have enough energy and those who do not
> but are increasingly willing to go out and get it. While
> nations have always competed for oil, it seems more and
> more likely that the race for a piece of the last big reserves
> of oil and natural gas will be the dominant geopolitical
> theme of the 21st century.
>
> Already we can see the outlines. China and Japan are
> scrapping over Siberia. In the Caspian Sea region, European,
> Russian, Chinese and American governments and oil
> companies are battling for a stake in the big oil fields of
> Kazakhstan and Azerbaijan. In Africa, the United States
> is building a network of military bases and diplomatic
> missions whose main goal is to protect American access
> to oilfields in volatile places such as Nigeria, Cameroon,
> Chad and tiny Sao Tome — and, as important, to deny
> that access to China and other thirsty superpowers.[28]

---

[27] "Profile: Hugo Chavez," in *BBC News International Edition* (Dec 3,
2007).
[28] Paul Roberts, "The Undeclared Oil War," *Washington Post,* June 28,
2004.

There are other shocks besides resource wars that can be attributed to fossil fuels, though. The immense volatility of fuel prices creates economic shocks that can drive the entire world's economies into recession on almost a moment's notice. The very recognition of that fact only tends to exacerbate the wild price swings of oil and other such commodities whenever fighting, or even the threat of fighting, breaks out in one of the world's major oil producing regions.

On a personal level, too, price swings affect people in very direct ways. When gasoline prices passed three dollars per gallon in the USA in 2006 (my apologies to all those in Europe and elsewhere who find such whining contemptible), sticker shock at the pump was all too serious for the working poor who had no other way to get to work than driving. Now, as this goes to print, gas prices are pushing $5/gallon. Unfortunately it is all too easy to direct one's rage at the seeming source of the problem, which demagogues are often happy to point out is the Middle East and its wealthy potentates (or that pesky Chavez). Never mind that generations of politicians have failed to create a mass transit infrastructure in the USA that could provide alternative modes of transport. We're a car nation, thank you very much.

Oil prices aren't the only thing to hit people hard in the pocketbook, however. Natural gas prices go as wild as oil, and heating bills have gotten so high sometimes that people end up shivering through winters trying to keep from going broke. It would be one thing if the supplies were actually as variable as price swings would lead one to believe, but there is ample evidence that crass manipulation of the energy markets is often more to blame than any actual supply shortfall. The most egregious example that comes to mind is the case of Enron, which cost consumers in the state of California many billions of dollars. But similar shenanigans have gone on for decades in both the oil and natural gas industries, and why not? The energy companies have insinuated themselves so deeply into the pockets of America's lawmakers (or vice versa) that they can be assured of nothing more than a slap on the wrist on the rare occasions when they're caught shaking down consumers. Meanwhile they rake in obscene billions in profits, with wars and unrest only serving to enhance their ill-gotten gains.

At the height of oil shocks, you could ask anybody at a gas pump if they'd like to be able to kiss OPEC goodbye if there was a realistic alternative, and their answer would be quite predictable. Of course the fossil fuel industry employs legions of workers, and the abandonment of an entire industry would have a serious impact. Yet the coal industry, in the course of a few decades, experienced a downsizing of some 90% of its workforce due in large part to automation and the closure of obsolete mines. The oil and gas industries—and what remains of the coal industry—are bound to pass into history as well. The march of progress makes the end of fossil fuel use an inevitability. How soon will such an energy revolution happen, and how fast will the transformation come to pass?

Poring over some of the thousands of articles in print and on the Internet, or listening to countless energy experts on television, one gets only a hazy impression of an elusively distant future when energy production and use will be transformed into a clean and affordable part of our lives. But the technology is not really the problem. Political will and the repudiation of the most powerful industrialists in the world are the main impediments to progress. A world of energy independence free of manipulation, and free of facile rationales for gouging consumers, is within our grasp in the immediate future.

Water Wars

The human population of the world stands today at about 6.7 billion. A great many of those people have difficulty obtaining sufficient fresh water for their needs. By mid-century the earth is expected to be home to some ten billion people. Where will all that extra fresh water come from?

This demographic horror story has resulted in predictions from many quarters of future wars being fought not just over energy supplies but over the most basic of human needs: water. Such wars have already been fought many times in the past, and international (and intranational) frictions that stop short of warfare are constantly at play around the world as populations struggle to appropriate water supplies sufficient for their needs.

Giant aquifers such as the Ogallala aquifer underlying several states in the middle of the USA are being pumped dry, far faster than their capacity to regenerate. Rivers are diverted for cities and irrigation, resulting in environmental catastrophes like the shrinking of the Aral Sea. It's sobering to imagine the pressures that will increase exponentially as earth's human population continues to expand, even as the glaciers that supply so many millions of people with their fresh water disappear under the relentless warming of the planet.

The deforestation and destruction of pristine habitat that is a corollary of overpopulation likewise destroys watersheds and further diminishes fresh water supplies. The impending water crises of the twenty-first century are as certain as the sun rising in the east, with the possible exception of massive disasters that would cull the human herd to more manageable numbers. With or without such catastrophes, things are looking pretty grim.

But don't give up hope. For the solutions to all these problems we've discussed—and more—are within our grasp, interwoven in a manner that may sound, at first, too good to be true. Yes, it will involve a paradigm shift and the boldness to embrace a global revolution. But it will be a joyful revolution, promising a more prosperous and peaceful world for everyone in the human family. Take heart. We're almost there.

CHAPTER TWO

Pie In The Sky

*Who can cloy the hungry edge of appetite
by bare imagination of a feast?*

— William Shakespeare, Richard II

VOICES FROM ALL sides are eagerly proffering solutions to the quandaries discussed in the previous chapter. Even the best of them, however, rarely attempt to fully remedy even a single one of these seemingly overwhelming challenges. Frequently the ideas are applicable to merely a portion of the world's population, usually those that are technologically more developed and which already possess substantial infrastructure for production and distribution of energy.

Unless one is unconvinced of the seriousness of global warming, nuclear proliferation, massive air pollution, nuclear waste, and political and economic instability caused by our dependence on fossil fuels, then it must be acknowledged that nibbling around the edges of these problems with half-hearted "solutions" is clearly insufficient. However well-intentioned they may be, virtually every proposal for addressing these urgent crises falls far short of its mark. Those who envision an environmentally benign technological utopia are usually, either intentionally or not, showing only half the cards in their hand, or badly misreading them.

In the previous chapter we only briefly touched on serious global problems that have already been the subject of numerous books, articles, and televised exposition and commentary. Since the intention of this book is to offer solutions, we will again be brief in discussing the remedies that are being proposed and how most fall regrettably short of even their modest goals. For those who pay close attention to these issues, much of this may not be new. But this background information is necessary to understand both the seriousness of the issues and the often deplorable shortcomings of their proposed fixes.

It is not my intention to question the earnestness or sincerity of those countless people who are attempting to analyze and solve some of the most pressing problems of our time. Yet it does a disservice to all to pretend that good intentions or limited goals will turn the tide. The global crises confronting us in the 21st century require solutions that will include everyone, from the most advanced city to the poorest village. This is not because of a question of fairness and social justice, though it would be wonderful if that was a sufficient incentive. The fact is that these environmental, political, and economic dilemmas already involve everyone in the world and cannot be solved except by solutions with global participation and applicability.

Most of the proposals that we will touch on here have merit, and are steps in the right direction. Added together, if we could implement many of them simultaneously, our situation would clearly be better than if we ignore the dire straits in which we find ourselves. But moving in the right direction isn't always sufficient, especially when the destination is far beyond the horizon. Sometimes we need a quantum leap, and this is one of those times. Yet since at least some of the proposals being bandied about purport to be The Big Answer, let's take a look at them and see if they're hiding an Achilles heel somewhere beneath their rosy scenarios.

Carbon Trading

This concept is so unutterably bogus that we should toss it on the slagheap right off the bat. In essence it is a deadly international shell game that allows corporations to buy the right to pollute in a great game of Environmental Risk. The futility of carbon trading can be inferred simply by observing that it's the one climate change amelioration scheme that seems agreeable to politicians and industrialists. An underdeveloped nation with a lot of trees but very little industry, for example, would rate as a carbon sink because of the carbon dioxide that its trees consume during photosynthesis. So that country could sell its unused polluting rights to some business in, oh, Dallas, Texas, for example. This unscrupulous hypothetical Texas utility company could then blithely belch out massive amounts of pollution from its coal-fired power plants via the simple expedient of buying the unused pollution rights from the poor yet tree-filled nation. Naturally its customers (and their neighbors downwind) foot the bill and reap the dire consequences.

It gets even more obscene than that, though. Developing nations like India and China, whose coal-fired plants just on the drawing board promise to vastly increase the blanket of global warming gases, are exempt from having to meet even the modest emissions targets under the terms of the Kyoto Accords. They can even sell polluting rights to the developed nations for every emission-reduction project they undertake. So, for instance, if China builds a hydro project, they can sell carbon credits thus earned to that imaginary(?) Texas company, in utter disregard of the fact that China's own ever-increasing fleet of dirty coal-fired plants is smoking away without a care in the world.[29] Now Texans too can have a little taste of Chinese air.

---

[29] Charles J. Hanley, "U.N. Nations Reach Deal to Cut Emissions," *Washington Post* Nov 17, 2006.

Carbon offsets are but one variation on this scheme. There are plenty of shady operations selling carbon offsets that don't really amount to anything but a con game to prey on those who would offset their carbon guilt.[30] The seller might guarantee that a certain stand of mature trees won't be logged, for instance, though the loggers will just go down the road to the next stand. Who's to say they won't come back and log the "saved trees" once the transaction is completed — or that any trees were ever saved at all? Billions of dollars are changing hands, with little recognition of the fact that carbon dioxide emissions are still pouring into the atmosphere. The myriad ways in which such systems can be gamed are limited only by the imagination of the shysters.

Carbon trading is little more than an unconscionable scam to further fossil fuel business as usual, and should not be considered to be any sort of real solution to the environmental problems we face. As Tom Burke, visiting professor at Imperial College London, has observed: "The reality is that applying cost-benefit analysis to questions such as [climate change] is junk economics... It is a vanity of economists to believe that all choices can be boiled down to calculations of monetary value."[31] Another commentator pointed out that carbon trading's "inherent complexity leaves it open to exploitation by special interests, not to mention perverse incentives to 'bank' pollution now against future credits."[32] This obscene ploy doesn't even deserve four paragraphs, but there you go. It's easy to find more information[33] on carbon trading if you'd care to explore it further, though on

---

[30] Fiona Harvey, "Beware the Carbon Offsetting Cowboys," *Financial Times* Apr 25, 2007.

[31] Kevin Smith, "'Obscenity' of Carbon Trading," in *BBC News International Edition* (Nov 9, 2006).

[32] Charles Komanoff, "Don't Trade Carbon, Tax It," in *Grist Environmental News & Commentary* (Feb 13, 2007).

[33] Smith, "'Obscenity' of Carbon Trading."

its face I trust that you, dear reader, can recognize a travesty when you see one. We have more serious ideas to discuss here.

Biofuels

Even for those who are vocal proponents of biofuels, a hint that there's something not quite copacetic about their lofty promise was presented by none other than George Bush Jr. in his 2006 State of the Union speech. In the up-is-down world of Bush policy — where the Clear Skies Initiative means that air pollution regulations will be relaxed and the Healthy Forests Initiative calls for more logging on federal lands — to hear him talk about the promise of ethanol sets off alarm bells in anyone who's been paying attention to his administration's appalling environmental record.

Ethanol is not the only biofuel, a general term that refers to a variety of fuels made from organic matter. The other main biofuel is biodiesel, in which vegetable oils of various types are modified, blended, or even burned directly in diesel-powered vehicles. Since plants soak up carbon dioxide as they grow, releasing that carbon dioxide when we burn them is essentially carbon-neutral, according to the most commonly understood explanation in their favor.[34] Both ethanol and biodiesel have serious problems, though, emblematic of a common fallacy that rears its head continuously in discussions of alternative energy.

Working in a laboratory, one can find elegant solutions to all sorts of problems. But extrapolating those solutions to global application is rarely feasible. It's like my neighbor who runs his diesel Mercedes on used vegetable oil from local restaurants. Works fine for him, but before you could provide enough such oil for all the cars on our block, let alone our whole town, we'd long since have run out of restaurants to supply it.

---

[34] L.J. Martin, "Carbon Neutral - What Does It Mean?," in *Ezine Articles* (Oct 26, 2006).

Obliviousness to the problem of scaling up tidy solutions to planetary size is a weakness of many alternative energy proposals.

Ethanol is a classic example. Discard for a moment the very real cost problem in creating even more blends of gasoline and ethanol, or the impact on the price of the sources of ethanol caused by hugely increased demand.[35] Just look at the amount of arable land that would be required to produce the ethanol needed to replace gasoline in the United States alone. The most optimistic figure I've seen for replacing all our gasoline consumption with ethanol calculates that we would have to double our cultivated land in order to meet the demand. If that's a best case scenario, then ethanol advocates must blanch when considering a study by the Worldwatch Institute indicating that to replace just 10% of transport fuel with biofuels in the United States would require 30% of its agricultural land. As bad as that sounds, it's not nearly as bad as Europe, which would require a staggering 72% of their agricultural land to produce biofuels for that paltry 10% figure.[36] Though globally the study's figures work out to 9%, the untenable figures for Europe and North America (Canada's is 36%) illustrate the drastic increase in land requirements as societies develop technologically and economically. Whereas a country like the USA might justify such a system as a reasonable use of excess corn capacity (up to now the production of ethanol from corn has received the most attention — and astronomical subsidies), the resulting slump in world corn supplies for food and the resultant higher prices to be borne by poor as well as rich nations raise serious ethical issues as well.

---

[35] Jeff Wilson, "Corn Rises as Argentina Halts Exports to Conserve Supplies," in *Bloomberg.com* (Nov 20, 2006).

[36] Fred Pearce, "Fuels Gold: Big Risks of the Biofuel Revolution," *New Scientist* Sep 25, 2006.

But the problem with corn-based ethanol is not only ethically questionable but economically ludicrous. In November of 2006 researchers at Iowa State University published an analysis that revealed that a price of $4.05 per bushel was the break-even point for corn prices, beyond which it would no longer make economic sense to build ethanol plants.[37] As I write this just five months later, corn future prices are hovering between $4.23 and $4.39 per bushel.

The ethanol industry in the United States is an artifice of government intervention. Without massive subsidies and tax breaks it would be just a fantasy. Direct ethanol subsidies costing about two billion dollars a year are a giveaway primarily to the corn industry, and Bush and other politicians who repeatedly extol the virtues of ethanol are simply greenwashing while fishing for votes from the corn belt. Okay, perhaps I'm being too harsh to accuse them of greenwashing, the deceptive practice of seeming to be environmentally friendly by touting green policies that the offenders know aren't viable. I have to admit that there has been ample evidence in recent years that many American politicians are abysmally ignorant (I will refrain, with difficulty, from naming names), so perhaps duplicity isn't always a factor. But make no mistake: greenwashing is a tactic widely used by both politicians and fossil fuel corporations.

Once he started his campaign for the 2008 presidential election, Senator John McCain suddenly became a supporter of ethanol subsidies when he realized he wanted to win the first caucus of the election season—in Iowa, the nation's corn capital. But here's what he had to say about the ethanol industry just a few years before:

---

[37] Tokgoz Elobeid, Hayes, Babcock, & Hart, "The Long-Run Impact of Corn-Based Ethanol on the Grain, Oilseed, and Livestock Sectors: A Preliminary Assessment," (Center for Agricultural and Rural Development, Iowa State University, Nov 2006).

"Ethanol is a product that would not exist if Congress didn't create an artificial market for it. No one would be willing to buy it," McCain said in November 2003. "Yet thanks to agricultural subsidies and ethanol producer subsidies, it is now a very big business — tens of billions of dollars that have enriched a handful of corporate interests — primarily one big corporation, ADM. Ethanol does nothing to reduce fuel consumption, nothing to increase our energy independence, nothing to improve air quality."[38]

Actually, Senator, it seems that ethanol actually worsens air quality. According to a report in *Environmental Science & Technology,* the 10,000 deaths in the USA annually attributed to pollution from gasoline engines may well get even worse with widespread use of the much-ballyhooed E85 (85% ethanol, 15% gas). A report out of Stanford University in 2007[39] described a study that predicted that with today's level of emissions, there could be up to 2.5 times more damage than the already considerable health toll of gasoline pollution alone.

Biofuel production has quickly created a global impact, with developing countries rushing to take advantage of the sudden thirst for ethanol and biodiesel in the USA and Europe. Brazil's soybean production has long been implicated in the destruction of its rainforests, even prior to the biodiesel boom.[40] Now burgeoning demand for biodiesel from soybeans has only exacerbated an already worrisome situation. Whereas logging gets most of the attention (it's easier to demonize chainsaws than tractors), look to the humble soybean and the vast

---

[38] John Birger, "Mccain's Farm Flip," *Fortune* Oct 31, 2006.

[39] "Biofuel's Dirty Little Secret," *New Scientist* Apr 21, 2007.

[40] Rhett A. Butler, "Soybeans May Worsen Drought in the Amazon Rainforest," in *Mongabay.com* (Apr 18, 2007).

quantities of them pouring out of the Amazon basin to find the reason behind much of the logging. Environmentalists feeling smug about eating tofu may have to rethink their situation. (In the interest of full disclosure, I consider myself an environmentalist and I eat tofu regularly. Mea culpa.) Soy, though, suddenly finds itself being upstaged. In 2007 the USA struck a deal with Brazil to clear even more forests, this time for sugar cane to make ethanol destined for the American market.[41]

When it comes to destroying rainforests in the service of biofuels, though, one would be hard-pressed to find a more devastating prospect than the developments in Southeast Asia:

Enter Malaysia and Indonesia, which together dominate the world market for palm oil. Palm produces significantly better yields of fuel per hectare than other crops. Both countries are now falling over themselves to increase production and, in late July, announced a joint plan to set aside 40 per cent of their palm oil output for biodiesel production.

Last year Indonesia, which already has 6 million hectares of palms for oil production, announced plans to expand this by 3 million hectares, partly by converting 1.8 million hectares of forest in Borneo - almost the size of Massachusetts - into what would be the world's largest palm oil plantation.

The expansion plan was condemned by Friends of the Earth and WWF [World Wildlife Fund]. The palm oil boom will "sound the death knell for the orangutan and hamper the fight against climate change, the very problem biofuels are supposed to help overcome," says Ed

[41] Tom Hirsch, "Brazilian Biofuels' Pulling Power," in BBC News International Edition (Mar 8, 2007).

Matthew, Friends of the Earth's palm oil campaigner. FoE claims palm oil plantations are the most significant cause of rainforest loss in Malaysia and Indonesia.[42]

Biodiesel suffers not only from land use issues but also from the inevitable combustion products of nitrogen oxides (which exacerbate smog and ozone problems), carbon monoxide, particulate matter, and carbon dioxide. And nitrous oxide is no laughing matter. Though manmade emissions of this gas amount to just 20% of global amounts, the fact that it's 300 times more potent in its greenhouse effects than carbon dioxide and that it persists in the atmosphere for over a hundred years should give us pause if we tend to be dismissive of that mere 20%.[43] We don't understand the feedback mechanisms well enough to casually assume that we can kick in an extra 20% without it making any difference. After all, in five years that's like adding a sixth year of planetary emissions.

Though biodiesel can be burned directly in diesel engines, most often the idea has been to blend it with mineral diesel. As with ethanol, biodiesel subsidies and tax breaks artificially prop up, at considerable taxpayer expense, an industry that promises, at best, to be only marginally better than just burning mineral diesel.

As grim as these brief observations serve to depict the realities of biofuels, there are many more negatives associated with them. Tad Patzek, an engineering professor at the University of California, Berkeley who's studied and written on the biofuel phenomenon, maintains that, "We're embarking on one of the most misguided public policy decisions to be made in recent

---

[42] Pearce, "Fuels Gold: Big Risks of the Biofuel Revolution."

[43] "Inventory of U.S. Greenhouse Gas Emissions and Sinks: 1990–2004," (US EPA, Apr 15,2006).

history."[44] Farmers in the middle of the USA who've been delighted to see the price of their corn rise as they invest in new ethanol plants are already beginning to have second thoughts, seriously concerned about the vast water demands that the process entails.[45]

The most deleterious impact, as is often the case, seems to be destined for developing countries. Whether it's clearing rainforest in Brazil to meet the USA's ethanol appetite or stealing land from small farmers in Columbia to plant palms for biodiesel,[46] we're seeing the ominous signs of a wholesale exploitation of land and people worldwide.

And what, pray tell, is to happen as the population continues to expand even as more and more land and water is dedicated to fueling the First World's vehicles? With the population predicted to rise by about 50% by mid-century, we are already faced with a formidable challenge for both food and water even in the absence of biofuel resource allocation. Even if we ignore the outrages already being perpetrated upon the developing countries in the name of biofuels, the misplaced priorities so clearly at odds with the coming population demands are staggering to contemplate.

Let us not be dismissive of the uncomfortable fact that billions of people subsist on substandard diets. It is to be fervently hoped that that situation could change for the better in the future, a not inconsiderable challenge when seen in the light of dramatically increasing population. Not only that, but as

---

[44] Nebraska College of Journalism & Mass Communications, "Most of Nebraska Corn Crop Will Go to Ethanol by 2011," in *News Net Nebraska* (Aug 30, 2006).

[45] Jim Paul, "Experts: Ethanol's Water Demands a Concern," in *MSNBC* (Jun 18, 2006).

[46] Tony Allen-Mills, "Biofuel Gangs Kill for Green Profits," *The Sunday Times* Jun 3, 2007.

societies increase in wealth they create a demand for more animal products, which puts an even greater strain on land use.

As the 2008 USA presidential election nears, it is distressing to hear candidates for that high office extol the promise of biofuels, even going so far as to paint a rosy picture of how demand for biofuels will provide jobs to poor people in Africa and elsewhere. Such outlandish views are directly at odds with the pleas from many organizations and individuals to call off biofuels programs for fear of the damage they will inflict on poorer, mostly tropical countries.[47]

> "We want food sovereignty, not biofuels...While Europeans maintain their lifestyle based on automobile culture, the population of Southern countries will have less and less land for food crops and will lose its food sovereignty...We are therefore appealing to the governments and people of the European Union countries to seek solutions that do not worsen the already dramatic social and environmental situation of the peoples of Latin America, Asia and Africa."[48]

The contention that biofuels represent a solution to our problems got a well-deserved splash of cold water by the Organization for Economic Cooperation and Development (OECD) in September of 2007. The study of biofuels by this highly respected organization that represents nearly every industrialized nation was presented in their report entitled *Biofuels: Is the Cure Worse Than the Disease?* Biofuel advocates would have a hard time responding to the OECD's scathing observations:

---

[47] Econexus, "Petition Calling for an Agrofuel Moratorium in the E.U.," (2006).

[48] Biofuelwatch, "This Is Not Clean Energy: The True Cost of Our Biofuels," (2007).

The rush to energy crops threatens to cause food shortages and damage to biodiversity with limited benefits ... Government policies supporting and protecting domestic production of biofuels are inefficient [and] not cost-effective ... The current push to expand the use of biofuels is creating unsustainable tensions that will disrupt markets without generating significant environmental benefits ... Governments should cease creating new mandates for biofuels and investigate ways to phase them out.[49]

In late 2007 the United Nations' independent expert on the right to food, Jean Ziegler, called for a five-year moratorium on biofuel production to halt what he described as a growing "catastrophe" for the poor. He called the increasing practice of converting food crops into biofuel "a crime against humanity," saying it is creating food shortages and price jumps that cause millions of poor people to go hungry.[50] It is distressing in the extreme that a system garnering such opprobrium from these respected organizations continues to be perpetuated by myopic politicians throughout the so-called civilized world.

One place where biofuels might prove to be practical, however, is in aviation. Commercial airliners operate within relatively narrow parameters when it comes to fuel, ever searching for fuels that will be energy dense, relatively compact and lightweight, and not prone to flash combustion in the event of a crash. Since they are depositing their exhaust directly into the stratosphere when at cruising altitudes, their GHG output is a concern even though the percentage of anthropogenic GHGs

---

[49] Richard Doornbosch and Ronald Steenblik, "Biofuels: Is the Cure Worse Than the Disease?," in *Round Table on Sustainable Development* (OECD, Sep 11-12, 2007).

[50] Edith M. Lederer, "Un Expert Seeks to Halt Biofuel Output," *Associated Press*, October 26, 2007.

contributed by aviation is but a small fraction of the total, only about 3%.[51]

The issue has not, however, escaped the attention of Sir Richard Branson, owner of Virgin Atlantic Airways. Besides putting his money where his mouth is by pledging billions to the search for global warming solutions, his Virgin Fuels initiative seeks to develop biofuels such as butanol to power airliners so as to make them carbon neutral. Expecting biofuels to power the world's automobiles may well be a pipe dream, but harnessing them to power the world's planes may be the best solution on the near horizon.

For those interested in further investigation, ample information on biofuel research is available with just a cursory Internet search. It is heartening to see that in the months since this section was first written there has been a considerable increase in the public recognition that biofuels pose serious problems. They may well serve a niche role in our energy future, but even the best-case scenarios for wholesale conversion to biofuels collapse under a little investigation. Besides, as we'll see later on in this book, there are much better sources for what limited biofuels we'll need than the places we're looking today. Those who are serious about global warming, air pollution, and fossil fuel dependency will have to do better than this. Economics and biology aren't the only issues here. There is a profound moral component to the equation that must not be ignored.

Clean Coal

When it comes to egregious polluters, the coal industry tops the list. Coal combustion is the world's foremost offender in the production of greenhouse gas emissions which, as

---

[51] "Inventory of U.S. Greenhouse Gas Emissions and Sinks: 1990–2004."

bad as they are, comprise only a portion of the coal stacks' damaging output. For years coal has provided the bulk of electricity generation around the world, with coal-fired power plants belching a toxic cocktail of pollutants. The resulting acid rain has killed forests, made lakes unin-habitable to fish, and dissolved ancient works of art where stone sculptures had the misfortune of being outside and downwind of power plants. Mercury and lead plumes have slowly but surely poisoned people and animals alike. Burn-ing over a billion tons of coal per year in the United States alone creates such stratospheric levels of global warming gases that the numbers make your eyes glaze over.[52] You may be trying to wrap your mind around that billion tons number, otherwise written as two *trillion* pounds. To bring that down to earth a little bit, that's about twenty pounds of coal per person — *every day!* Add up the coal consump-tion of China (horrendous), India, Europe, et al, and you're talking about some serious pollution. If you happen to be a global warming nonbeliever, just how would you imagine that such an amount of smoke could be poured out without affecting our planet?

Yet coal is cheap and plentiful, and in the face of supply and price volatility of other fuels, fears of global warming aren't stopping governments and utility companies from building even more coal-burning power plants to fill ever-increasing de-mands for energy. In the USA there are 154 coal-burning power plants on the drawing board in 42 states.[53] China is building new coal-fired power plants at the rate of about one large plant per week, this despite the fact that 16 of the 20 most polluted

---

[52] Kentucky Educational Television, *KET Coal Mine Field Trip Q & A* (Dec 15, 2005 [cited); available from http://www.ket.org/Trips/coal/qa.html.

[53] Dave Hoopman, "Home-Grown Energy Sources Continue to Expand," in *Wisconsin Energy Cooperative News* (Jan 2007).

cities in the world are in China.[54] But countries need energy, and sitting on such vast energy reserves has caused politicians to ignore the consequences of their coal-burning folly in favor of the cheap shortsighted fix. Yet new technologies have been developed to actually make coal a considerably cleaner fuel. Unfortunately, the cheaper dirty coal power plants continue to be built around the world.

Considering the vast amount of environmental damage already done by coal burning, and the even more serious damage virtually foreordained by the building of so many new plants, there is an almost overwhelming urge to dismiss talk of "clean coal" and low-emission coal power plants as unrealistic. Dan Becker, director of the Sierra Club's Global Warming and Energy Program, states, "There is no such thing as 'clean coal' and there never will be. It's an oxymoron."[55] And Green Scissors, an environmental coalition, claims that coal can never clean up its act. "Because of the basic chemical and physical characteristics of coal," the group says, "once [it] is burned, the reduction of CO_2 emissions becomes economically impossible."[56]

One would think that blanket statements of what's economically impossible when it comes to technology would be more cautiously asserted. In point of fact, new technologies exist that hold out the possibility of coal-burning power plants with dramatically reduced GHG emissions. This depends, of course, on all the elements working as predicted. The new plants would utilize what's called "coal gasification," in which the coal is first turned into a gas that is then cleaned before it's burned. The technical name for the system is Integrated Gas-

---

[54] CBS Evening News, "The Most Polluted Places on Earth," (June 6, 2007).

[55] Amanda Griscom Little, "Coal Position," *Grist Environmental News & Commentary* (Dec 3, 2004).

[56] Lisa Kosanovic, "Clean Coal? New Technologies Reduce Emissions, but Sharp Criticisms Persist," *E: The Environmental Magazine* Jan-Feb 2002.

ification Combined Cycle (IGCC). Sulfur, mercury, lead, and carbon dioxide can be removed from the gas before it's burned. The system promises to be a vast improvement over current pulverized coal power plants, utilizing about 40% as much water as current coal plants and producing about half as much solid waste (which is still full of nasty pollutants and has to be disposed of somewhere, however).

The carbon dioxide from such plants can be collected instead of sending it up the smokestack. At that point another new technology would be used: carbon sequestration. The carbon dioxide would be compressed and injected deep underground, where theoretically it will remain for thousands of years. Actually carbon dioxide has already been used in this way on a much smaller scale in order to coax more oil out of underground reserves. Estimates of the storage capacity available deep underground usually tend to indicate that there is sufficient space for over a hundred years' worth of coal burning. Of course no one has ever attempted to pump billions of tons of carbon dioxide underground before, and naturally there are fears that some of it may leak back to the surface and thus into the atmosphere. Pilot projects, however, and prior experience seem to lend credence to those who are promoting this system. The Intergovernmental Panel on Climate Change estimated in 2005 that in excess of 99% of carbon sequestered is "very likely" to remain in place for at least one hundred years.[57]

Here again is that weighty clause "very likely." While not meaning to sound like a Cassandra, it might nevertheless be worth pointing out that a massive planetary belch of carbon dioxide would hardly be simply a question of its impact on the atmosphere. One evening in 1986 a cloudy mixture of carbon dioxide and water droplets erupted without warning from the

---

[57] IPCC, "Underground Geological Storage," in *IPCC Special Report on Carbon Capture & Storage* (Sep 2005).

depths of Lake Nyos in Cameroon.[58] As the ground-hugging mist[59] dispersed through nearby villages for over 20 kilometers in all directions, it killed over 1,700 people, as well as most of the animals in its path. Just two years earlier a similar occurrence had killed 37 people at a nearby lake. How comfortable do you feel about the assumption that billions of cubic meters of carbon dioxide will "very likely" remain underground if we decide to pump it down there? Ever heard of cracks and fissures in the earth's crust? Earthquakes? Would you want it anywhere near where you live?

Even if we assume a best case scenario, could this really be a solution? The U.S. Department of Energy (henceforth the DOE) has sunk a cool billion dollars into a prototype clean coal power plant called FutureGen.

> Proposed in 2003 and backed by a consortium of coal and electric companies, it is not due to come online until at least 2013. Many in the industry consider this date to be dubious, nicknaming the project NeverGen. [In fact, as this book was being edited, FutureGen was indeed canceled.] It is intended to make it look like the coal industry is doing something, while actually doing very little and in the process putting off changing how coal plants are built for a decade or two. [According to] its Coal Vision report, the industry does not plan on building "ultra-low emissions" plants on a commercial scale until between 2025 and 2035. According to the report *there is considerable debate about the need to reduce CO₂ emissions.* [My outraged italics!] The report also states that "achieving

---

[58] US Geological Survey, *Volcanic Lakes and Gas Releases* (1999 [cited]); available from http://tinyurl.com/4fd7fa.

[59] Carbon dioxide is about 1.5 times as dense as air, a critical feature of its hazardousness.

meaningful CO_2 reductions would require significant technical advances."[60]

No wonder environmental groups are skeptical of the coal industry. One would be hard pressed to come up with a more callous case of greenwashing. Even if all the technologies for clean coal energy generation work as advertised (as they seem to in pilot projects), and even if we're content to have mountain tops ripped off and dumped into river valleys and all the other horrific environmental damage done by the extraction of coal, the promise of clean coal will only be a mirage since it won't be implemented for decades. But we can absolutely not afford to wait.

> A window of opportunity for clean coal is closing, advocates warn. If populous nations such as China and India keep belching out coal emissions as they chase prosperity, "we can pretty much wipe out any chance of dealing with global warming in this century," said John Thompson, advocacy coordinator for the Clean Air Task Force.[61]

And it's not only China and India who are the offenders. Utilities across the United States are rushing to build dirty coal plants before new environmental regulations force them to clean up their act. Once all the plants now on the drawing board come online, they stand to increase atmospheric CO_2 levels up to four times their pre-industrial levels.[62]

---

[60] James, *Clean Coal or Dirty Coal?* (Oct 3, 2006 [cited]); available from http://tinyurl.com/3wxh39.

[61] Kevin Coughlin, "King Coal Comes Clean," *The Star-Ledger* Mar 6, 2005.

[62] "Climate Change Post-2100: What Are the Implications of Continued Greenhouse Gas Buildup?," in *Environmental and Energy Study Institute (EESI)*, U.S. Congress (Washington D.C.: Sep 21, 2004).

For anyone who isn't in abject denial about global warming, this situation should make them shudder with foreboding.

You'll hear coal advocates extolling the virtues of carbon sequestration as if they've got their eye on the holy grail. But even if they actually implement that technology and it does work as advertised, let's not forget that there is an incredible amount of ash that remains after the coal is burnt, laden with everything from lead to mercury to uranium. Up to now that ash has been dumped fairly haphazardly without much regulation or concern for its impact on the biosphere. When you take all the negatives into consideration, "clean coal" definitely looks like an oxymoron. And according to even the most vocal advocates of such a plan, it's going to be very expensive.

Clean coal advocates do their best to convince people that sequestering carbon dioxide is all that needs to be done. Yet greenhouse gases are released during and after the mining process, gases that have no way to be sequestered because they begin to escape into the air as soon as the overburden is stripped away in opencast coal mines. Much of what is uncovered is carbon-rich shale and mudstone, and the methane, carbon dioxide and carbon monoxide that they contain will continue to be released into the atmosphere. Amounts vary depending on the mine, but anyone who tells you that greenhouse gases are no problem if you use clean coal technology and carbon sequestration is not to be trusted. In fact, many coal seams contain so much methane that they are tapped for their methane rather than their coal.[63] But those coal mines in which the methane is less concentrated simply release their often considerable quantities of methane into the air. Nearly 10 per-

---

[63] R.M. Flores and L.R. Bader, "A Summary of Tertiary Coal Resources of the Raton Basin, Colorado and New Mexico," in *U.S. Geological Survey Professional Paper 1625-A* (U.S. Geological Survey, 1999).

cent of atmospheric methane resulting from human activity is derived from coal mining.[64]

In order to prevent an environmental catastrophe due in large part to coal burning, the governments of the world will have to not only come up with a viable near-term solution to clean energy generation. They'll actually have to cancel their planned coal plants and decommission the ones they've already built. Yet the technology, as we'll see later, will be the least of our problems. The politics, on the other hand... Aye, there's the rub.

Natural Gas

The past couple of decades have been a boom time for natural gas, as more and more utility companies have placed their bets on it as an alternative to coal. Politicians of both major parties in the U.S. have waxed enthusiastic about building another lengthy pipeline across Alaska and even down through Canada to the lower 48 states, extolling the virtues of natural gas as clean and environmentally friendly. But is it?

> This is a half-truth at best. Methane-rich natural gas is a much more dangerous greenhouse gas than carbon dioxide. [Methane is 20 times more potent in its effect] Already, about 2.3 percent of the natural gas produced by the industry leaks out of valves, pipes and other infrastructure, unburned. If that proportion makes it up to 3 percent, using natural gas is no better for the atmosphere than burning oil.[65]

---

[64] "Coalbed Methane--an Untapped Energy Resource and an Environmental Concern," ed. U.S. Geological Survey, Energy Resource Surveys Program (Jan 17, 1997).

[65] Sonia Shah, "The End of Oil? Guess Again," in *Salon.com* (Sep 15, 2004).

Granted, natural gas avoids some of coal's nasty emissions, but relying on it in the long term is no solution, since even if you could stop the inevitable leakage it still produces prodigious amounts of greenhouse gases as it burns. Given a choice between coal and natural gas, the latter would be the obvious choice from a pollution standpoint, though its price volatility has made it a real concern for those who built gas generators only to find the price jumping due to the increased demand. Clearly natural gas is a stopgap measure that, at best, can fill in as coal is abandoned. But both of them, as well as their fossil fuel cousins, have to go, and the sooner the better.

Energy Efficiency

The insatiable electrical demand of the United States increased over the last thirty years, on a per capita basis, by 50%. One exception to this formidable increase was California, where demand stayed flat.[66] What is it about California that could cause such a marked difference in demand? Air conditioned homes are ubiquitous, temperatures are all over the map, yet even with all the technological goodies that often make their debut in California its residents just sip electricity compared to most of their fellow Americans.

The difference is the state's enthusiastic support for energy efficiency programs. Since the economics of spending money to promote energy efficiency clearly show an advantage over building new power plants, the states' utility companies were enlisted to coordinate all sorts of programs to promote saving electricity. Many of the incentives involved simple rebates on everything from freezers to light bulbs. Californians could get a bounty, of sorts, for turning in old appliances that were grossly inefficient. When compact fluorescents first came on the scene

---

[66] N.Z. Electricity Commission, "Electricity Efficiency Can Influence Future Load Growth," (Sep 5, 2005).

at high prices, they could often be bought for a dollar apiece (and still can be) at local hardware stores, subsidized by the power companies (i.e. the public consumers of electricity, via the power companies).

In most cities and towns across California, incandescent traffic lights were converted to LED technology. It may sound like a small thing in and of itself, but imagine how many streetlights there are in California. To take one example, the city of Gardena, in Los Angeles County, has a population of about 57,000. In 2002 the city installed 1,688 LED traffic lights at a cost of about $540,000. The city estimates that it saves about $109,000 per year in electricity costs as a result, meaning that in five years the savings more than make up for the installation.[67] Elsewhere in California, the city of Hanford has had LED traffic lights for about a decade, and is just now starting to see the need for some replacements. Assuming that Gardena's lights yield the same longevity, the electricity savings will have paid for both the original installation and full replacement by the time they start to fail, with enough left over to throw a party for the city officials who were wise enough to take the leap.

The dramatic example of California's energy efficiency efforts, still far less than what could easily be accomplished, clearly points the way. If that state's modest programs were to be extended nationwide, electricity demand in most of the other 49 states would drop by approximately one third. Projecting such programs worldwide would accomplish energy savings that would be truly staggering.

A ban on the manufacture of incandescent bulbs, by itself, would be a huge leap. As this is written there is a bill pending in the California statehouse to enact just such a ban. It has been calculated that replacing four 100-watt incandescent bulbs with

---

[67] "California Says "Go" To Energy-Saving Traffic Lights," (US DOE, May 2004).

their equivalent in compact fluorescent lights (CFLs) in each household across the USA would save energy equivalent to the output of more than a dozen 1-gigawatt (GW) power plants. If incandescent bulbs were banned outright, those savings would be considerably greater. Projected worldwide, it is entirely believable that such a ban on incandescents could replace many dozens of power plants.

Compact fluorescents have come a long way in the past twenty years, and LED technology is rapidly approaching home lighting applicability, already finding uses in directed lighting fixtures. But it gets even better. Cold cathode bulbs (CCL), already in use in some applications and making great strides in development, are even more efficient. CCLs produce about 25%-50% more light per watt than even CFLs, last about four times as long, can be used with either alternating or direct current, and have full dimming capability and immunity from damage due to voltage fluctuations. While currently more expensive than CFLs, the disappearance of incandescents would stimulate the market for all types of energy saving lighting and undoubtedly lead to further refinements, as well as price reductions due to economies of scale as mass production ramps up.

Replacing incandescents around the world with CCLs would save even more energy than using CFLs. Already they are made in both tubular and screw-in bulb configurations for easy replacement of both incandescents and standard fluorescents. Cold cathode lamps are the most promising candidates for utilization in off-grid applications such as single-dwelling solar applications in Third World countries. But cold cathode lights would save considerably more than meets the eye because of their dimming ability. Anyone who uses dimmers in their home knows that much of the time the lights are kept below full brightness. When a cold cathode bulb is dimmed even slightly its energy consumption drops substantially. Thus the obvious efficiency advantage of 25-50% over the relatively frugal CFLs would translate into even greater savings.

Cold cathode bulbs have been around for some time, since their ruggedness and low power consumption make them ideal for mobile applications like boats and recreational vehicles. With a life span of 18,000-25,000 hours, they end up outliving the vehicles they're used in. One would think that they would be the bulbs to rave about, yet they're still an oddity unknown to most Americans. Part of the reason is that the brightest ones to be deployed in standard screw-in shapes are still only about as bright as a 60-watt bulb, which even so makes them ideal for many applications. Might another part of the reason be the fact that they would last nearly a lifetime?

When Walmart decided to sell a hundred million compact fluorescents they approached General Electric to partner with them. GE balked, since their profits from the sale of ten incandescent bulbs would far exceed what they would make selling one CFL that would last about as long. Besides, they have all the manufacturing capacity to build incandescents. No matter, said Walmart, we're going there whether you're on board or not. GE had no reasonable choice and knuckled under to Walmart's demands.

It's highly unlikely that Walmart was unaware of cold cathode fluorescent technology. Why, then, didn't they promote that, at least for the large portion of their program selling 60W equivalent bulbs or less? GE, after all, was being forced to retool to supply massive volumes of CFLs. Why not have them instead retool for CCL production? Could it be that CCLs are too good for Walmart? After all, Walmart is a corporation that has to make a profit too. Maybe when they looked at the numbers they realized that the bottom line for CCLs looked pretty skimpy.

CCLs have an expected lifetime, as mentioned above, of 18-25,000 hours. If you calculate the usable life of a 20,000 hour bulb that's burned for four hours a day — most bulbs in a household are used much less than that — it means that one bulb would last over 13 years. If you swapped the most-used bulbs in your house with those used least every five years or so, you could reasonably expect that a household full of CCLs would last at least 25 years! Their durability would even protect them against many knockover accidents that would break an incandescent. The prospect of selling light bulbs to their customers once every generation or so probably didn't look anything like a sound business decision to Walmart.

This is a classic example of capitalism colliding with principle, which we'll see more of in the pages to come. Walmart can well be applauded for pushing through their CFL program, and from a corporate economics standpoint can hardly be blamed for not taking it a step farther. On the other hand, CCLs are virtually certain to become more widely known (look, now YOU know about them!), and where there is demand there will eventually be someone who will supply them. Does it make sense to have GE and others building CFL production facilities for a technology that is practically already obsolete? Given the seriousness of global warming pressures, wouldn't it make sense to do the best we can when it comes to saving energy?[68] LED technology is making great strides as well, and many within the lighting industry expect them to be vying for the home lighting customer in the very near future. They light with no flicker, consume very little electricity, and are so durable that they'd last the life of the house. It seems strange to be gearing up production for CFLs worldwide when we can already hear the opening notes of their swan song.

Another no-brainer energy saver is a tankless water heater. It's well known that electric tank-type water heaters are more expensive to operate than similar gas heaters. Yet an electric tankless water heater costs about half as much to operate as a

---

[68] Full disclosure: I have no financial interest in CCL manufacturers, or any other lighting companies.

gas tank-type heater. If you've ever had a tankless heater you'll have experienced the humble joy of knowing you'll never run out of hot water, even if you've got a houseful of guests who just came back from a mud-wrestling competition. They're simple to install, take up much less room than a tank heater, and last about twice as long. Yes, the initial cost is a bit more, but the energy you'll save during their lifetime more than makes up for the price differential.[69] Would it make sense for governments to publicize this fact, impose a modest tax on tank-type heaters and use it to subsidize tankless heaters in order to encourage their widespread adoption? I know the libertarians and small-government fanatics will be gnashing their teeth over ideas like this, but are we serious about saving energy or not? If Grover Norquist is okay with having a coal-fired power plant in his backyard, I'll listen to his complaints about such social engineering. If not, he and his ilk can keep their big mouths shut.

If the bottom line is what's so important, here's a good bottom line for you: Proven energy efficiency programs can eliminate the need for hundreds of power plants, and we can enact them right now. If corporations are responsible to their stockholders and too often guilty of thinking only of short-term results, governments should be their counterweight, responsible to the citizenry and considering the long-term benefits of their policies. It's about time we held our policymakers' feet to the fire. There is absolutely no excuse to avoid sweeping energy efficiency legislation, and no question whatsoever about the results.

Electric Cars

The summer of 2006 saw the release of a scathing documentary entitled *Who Killed the Electric Car?* The movie revealed

---

[69] SEISCO, "Comparison of Estimated Annual Water Heating Costs," (2007).

the guilty parties behind the creation and ultimate forcible destruction of all-electric automobiles produced to meet California's mandate of zero-emission vehicles. The prospect of a revolutionary new vehicle technology being accepted by the public suddenly rallied the fossil fuel companies, auto manufacturers, and compliant government officials to short-circuit a budding transformation.

Part of the threat to the auto industry that the electric car represented was a result of its simplicity. Most of the parts and systems inherent in every internal combustion vehicle were eliminated, impacting industries worth billions of dollars and cutting into the profits of automakers. No corporation intends to roll over and blow away without a fight, much less a large number of interrelated corporations that rely on a transportation industry that's developed symbiotically over the past hundred years.

The same sort of stultifying pressure from the powers that be can be expected for any truly revolutionary propulsion system that threatens to upset the fossil fuel applecart. Yet upset it we must. *Who Killed the Electric Car?* is a cautionary tale for anyone who's really serious about transforming the energy infrastructure. As difficult as it was to make even this small inroad into the automobile/petroleum colossus in the United States, it pales in comparison to what must be done globally. Automobiles are but one facet of the revolution to come.

The promise of the electric car masked an ugly truth, however. For while its owners tooled around emitting no exhaust, feeling smugly eco-correct, few seemed to be concerned about the source of the electricity that powered their vehicles. They spoke glibly about saving the planet, yet the electricity used to charge up their cars day in and day out was most likely being generated by coal-fired power plants. The pollution was happening farther away, but the EV-1 was far from being environmentally benign. Out of sight, out of mind. But if you had to watch the coal smoke pouring into the sky, would you feel as self-righteous about your electric car?

As this was being edited in 2007, a company called Altair-nano claimed to be ready to introduce a new type of battery in partnership with a startup electric car company called Phoenix Motorcars. Preparing to go to press in 2008, however, there's still no sign of it. Supposedly this battery will have long life, a high safety factor, and the ability to charge for up to 250 miles in a mere ten minutes.[70] If this is true, it represents a huge paradigm shift in automotive transportation, with ominous implications for the oil industry and wonderful implications for the air in our cities. Provided, of course, that those cities aren't downwind from the many new power plants that will have to be built to provide the electricity to run them.

Non-polluting renewable energy sources produce but a small sliver of the electricity consumed every day. Until our primary energy sources are clean, electric cars and their kind will still be inextricably tied to environmental degradation. Once we do deploy clean primary energy sources, though, electric cars will be able to come out of the shadows again, better than ever for the intervening improvements in technology.

Solar Power

The sight of a photovoltaic panel with an electric meter spinning quietly at its base is a powerful and reassuring image to almost anyone who longs for a sane energy future. Indeed, for many solar proselytes, gazing at a field full of solar collectors is an almost spiritual experience. It represents not only wonderfully clean energy technology but a worldview considerably more desirable than those espoused by the

---

[70] It should be noted that this charge rate would take a very high-energy connection at specially equipped charging stations. Don't try this at home!

politicians and plutocrats who are in charge of energy policy today. Just scale that up a bazillion times and we're in energy utopia.

It is hard to deny the solar aficionado's contention that one reason it hasn't made any more headway than it has (and it is making headway) is because solar's been starved for R&D funding. I certainly won't try to refute that here. Solar and wind power research were on the short end of the funding stick for decades. That has changed in recent years, however, but even today, much of the progress we see is due to entrepreneurs and academic institutions.

So just how rosy is the solar picture? It's certainly getting in the news more these days, though considering the tsunami of articles, studies, and alarm about global warming you'd think solar would be even more front and center. Some major projects are being implemented, though, and others that are already in place seem to be operating as well as expected.

There are basically two types of solar energy technologies when talking about electrical generation: photovoltaics (PV) and heat concentrators. Passive solar for heating purposes is a separate field in its own right. Photovoltaics convert sunlight directly into electricity using solar cells. Heat concentration schemes, on the other hand, focus sunlight using curved mirrors to heat a liquid that is then utilized via a heat exchanger and turbine to generate electricity.

The latter system seems to be more efficient than PV cells at this time, and also cheaper. Nevertheless, solar-generated electricity is still considerably more expensive than that generated by the major players (coal, gas, nuclear, and hydro). Any builder of these systems, however, will be quick to point out that economies of scale will surely diminish the costs once wider deployment and its attendant mass production kick in, and that's undoubtedly true. Individual homeowners buying PV panels are still fairly rare, for the payback period is very lengthy at current prices, and if you're off the grid you still need pretty substantial backup systems of batteries and inverters.

Just what is the payback time for photovoltaics? Proponents of such systems are often either quite secretive about their actual efficiencies or considerably overoptimistic. But the Sacramento Municipal Utility District has had some direct experience with a large array of photovoltaic panels that fed their system, and they divulged some efficiency figures that shed a little light on the numbers:

> "The Sacramento Municipal Utility District (SMUD) did slip me some performance numbers. PV systems produce about 1,400 kWh per year for each installed one kWe Solar PV system. Thus the capacity factor of this solar PV system is only 15%. No system with capacity factors this low is a viable energy producing system."[71]

Well, actually that'd be almost 16%, but it's almost academic. A typical installed 2kW PV system in 2007 costs about $18,000.[72] (The graph a few pages after this lists solar costs as a little under $6,000/kW rather than this $9,000, since it takes into account the lower cost per kW for commercial massed arrays which could comprise a substantial segment of solar electric production.) Even though most residential roofs in the USA have some shade (most homes have trees planted quite purposely to shade the house during the hottest part of the day, for obvious reasons), we'll assume that not only is there no shade at all but that the solar intensity is equal to the relatively high values in the Sacramento valley.

---

[71] P. E. Donald E. Lutz, "PG&E Solar Plants in the Desert," in *Truth About Energy* (2007).

[72] Solarbuzz Research & Consultancy, *Solar Photovoltaic Electricity Price Index* (2007 [cited]); available from http://www.solarbuzz.com/SolarIndices.htm.

Given those benefits of the doubt we can estimate an annual output for this 2kW system of 2,800 kWh per year. The average cost of electricity from the U.S. grid as of March 2006 was 9.86¢/kWh.[73] At this rate the 2kW installation would produce $276 worth of electricity per year. The payback rate for the installation (generously and unrealistically not counting interest on any loans or maintenance or replacement of equipment during that time) thus comes to a bit over 65 years. In addition to that untenable prospect, if the home uses the USA average of about 888 kWh per month,[74] the additional electric bill would add an annual expense of $774 to the family budget. Of course many homes and businesses rely on hefty subsidies to offset these clearly unworkable economics, but someone has to pay for the real cost down the line, and that someone is you and me. Without massive subsidization, the PV industry would simply not survive except among the very wealthy and those living off the grid with very limited options.

Most of the buzz today surrounds big solar projects where fields of solar arrays produce a substantial amount of power that is then distributed via the grid. There's a new system being built right now in Nevada called Nevada Solar One that uses a trough design of solar concentrator and is scheduled to produce 64 megawatts (MW)[75] in a field of shiny reflectors covering some 350 acres, a bit over half a square mile. For comparison's

---

[73] Michael Bluejay, *Saving Electricity* (2007 [cited); available from http://tinyurl.com/2nks3f.

[74] Energy Information Administration, "Residential Consumption of Electricity by End Use, 2001," (DOE, 2001).

[75] A megawatt = 1000 kilowatts = 1,000,000 watts. Sometimes this is written MWt to denote thermal energy (as in a power plant) or MWe to denote the electrical output, necessarily lower because of the conversion of heat to electricity in the turbine system. For purposes of simplicity in this book I will use simply MW to always denote MWe, the electrical output or consumption in question.

sake, a coal-fired power plant produces about 1000 MW and nuclear plants perhaps 1,300 (though nuclear plants are often clustered, as in France where they put four of them together for a 5-6,000 MW total output).

That "half a square mile" figure bears a bit of looking into because solar proponents sometimes get quite disingenuous when it comes to their acreage figures. Half a square mile would be, in a typical configuration, a rectangle half a mile on the short sides but a mile long on the long sides. "Half a mile square" would be half that area, a literal square with each side being a half-mile. So "miles square" vs. "square miles" in this instance yields a smaller area. Since area is almost always discussed in terms of square miles, though, the disingenuousness kicks into high gear—and in the opposite direction—when you start scaling up the solar arrays.

Here are a couple of quotes from an article on what is scheduled to be the largest solar energy "farm" yet, a 4,500 acre, 500 MW array of Stirling solar-thermal dish generators in the Mojave Desert in southern California:

> [...D. Bruce Osborn, Stirling Energy's new CEO, says,] "a dish farm of 11 miles square could produce as much electricity as the 2,050 MW from Hoover Dam."
> ...Theoretically, Stirling dish farms with a total area of 100 miles square could replace all the fossil fuels now burned to generate electricity in the entire U.S.[76]

A casual reader might be forgiven for interpreting that "11 miles square" figure as the more usual "11 square miles," but in reality it denotes 121 square miles. That is a LOT of dishes. It's also a wildly inaccurate calculation, since at a capacity of 500 MW per 4,500 acres it would take approxi-

---

[76] Otis Port, "Power from the Sunbaked Desert," *Business Week* Sep 12, 2005.

mately 29 square miles of dishes — not 121 — to equal that 2,050 MW Hoover Dam output. While this may look good at first glance, that would be around 80,000 dishes (each 37 feet in diameter), which will soon allegedly cost $150,000 each (they cost much more than that now) and could, we are told, drop to half that cost with true mass production. Not to get bogged down in figures here, but even at the theoretical greatly reduced target figure of $75,000 each that would come to a tab of about six *billion* dollars. Not exactly chump change.

The second figure quoted above is pretty close to right on (it seems the reporter is better at math than the solar guy), if by fossil fuels you also (incorrectly) include nuclear power. But 100 square miles isn't what they're talking about there. Notice it's "100 miles square" which is actually 10,000 square miles, an area larger than the state of Vermont. Oh heck, as long as I've got my calculator out I'll do the math for you. That would take about 28 million dishes and run up a bill of around $2.14 *trillion.* Plus the cost of all the feather dusters you'd need to keep them clean and shiny.

If that 10,000 square miles sounds like a lot (and it is!), an article in the esteemed journal *Science*[77] by a proponent of solar concentrator technology, as exemplified by the aforementioned Nevada Solar One, estimated that to supply 50% of the USA's *present* energy requirements would require 15,000 square miles of solar panels in the desert southwest. Not to be outdone, *Scientific American* touted a plan to provide 69% of America's electrical needs by 2050 with a plan to cover 30,000 square miles with solar panels![78] Construction of such a system would require completely covering 2 square miles per day

[77] Reuel Shinnar and Francesco Citro, "A Road Map to U.S. Decarbonization," *Science* 313 (Sept 1, 2006).

[78] James Mason and Vasilis Fthenakis Ken Zweibel, "A Solar Grand Plan," Scientific American January 2008.

with solar panels and all their supporting infrastructure, every single day for over forty years. One can't help but wonder at the limitless imagination of those who propose such scenarios with seemingly no thought for the implications of scaling up construction projects to such unrealistic sizes.

Upgrades to the transmission grid would add another $1.1–1.3 *trillion* to the already staggering cost of building such installations. The estimate for line loss of about 7% seems unrealistically optimistic considering how far the electricity would have to travel under such a scenario, especially since 7% is a tad less than we lose with our current grid, and today we don't try to push the juice nearly that far. Maintenance, including repeated cleaning of the solar concentrators, was not mentioned anywhere. That can hardly be dismissed as inconsequential, considering that the 15,000 square mile solar array proposed in *Science* would have to cover the equivalent of the entire states of both Connecticut and Vermont.

Of course you'd still need a backup system from dusk to dawn, and since all the solar arrays couldn't be in the sunniest location the efficiency would be considerably less overall, but let's not even go there, okay? Just remember that we in the USA, who use a prodigious amount of electricity per head, aren't the only ones living on this planet. Yes, I know that's a shock, but revelations like that are why it's good to read a book once in awhile. And pretty much everybody wants electricity. Are we serious about global warming? Energy wars? Air pollution? If the 5% of the world's population that lives in the USA is willing and able to come up with at least two trillion and change to go solar, all we'll have to do is convince the other 95% to do the same. Not gonna happen.

Oh, you can quibble about the figures a bit, but the Stirling array discussed above—and its optimistic cost projections—were chosen as a bellwether demonstration of solar feasibility based on Sandia Laboratories' Solar Thermal Test Facility. At the rate they've agreed to sell their power to California utilities it'll take about 67 years to pay it off (not counting interest, maintenance, upkeep, and replacement costs for failed components). The Nevada Solar One trough system costs about the same amount, around $3 million/megawatt (and lets not forget the $1.1+ trillion for grid upgrades). If you want to talk seriously about energy costs and renewables, you might want to take a peek at this chart before you dump all your eggs into the solar basket:

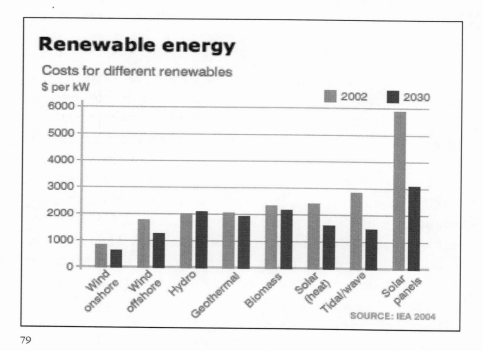

79

My comment about feather duster costs was not entirely tongue in cheek. Keeping solar panels clean so that they run as efficiently as possible, whether PV or reflectors, requires frequent pressure washing — every ten to twenty days, for example, at a large trough reflector system in Cali-

---

79 IEA, "Costs for Different Renewables," (BBC News, 2004).

fornia. Solar arrays are generally sited in deserts to avoid cloud cover. Where will all the water come from? The cost of piping or trucking in all that water for seven decades or so should really be factored in to get an accurate picture of a solar array's electricity price. How much will water cost in the desert fifty years from now? More than today — you can bet on it.

However much solar power cheerleaders try to blame a lack of effort and funding, the problem with widespread use of solar power for electricity is not a question of either apathy or conspiracy. It's simply a matter of physics:

> The reason is simple. Solar energy is dilute. Once it's collected the various applications become possible. But to collect it in the amounts required to make a real difference is a huge difficulty. There is no short cut, no technology can be invented to surmount it: massive areas of the earth's surface must be devoted to it. Solar energy has been well understood for over a century; the amount of solar energy falling on the planet is known, fixed and unchanging. The areas required for collectors, if solar was to make a significant contribution on the scale of present energy needs, are, in turn, on the scale of entire states.
>
> Efficiency increases to the limit the physics allows do not alter the issue. The scientific and engineering realities are plain. The amounts of materials, even cheap materials, the land areas occupied, the maintenance required, and also, more than possibly, the lawsuits brought by the very environmental industry promoting solar, make the whole solar enterprise on the scale required to power the nation a dream, not a practical reality, not now, not in the future.[80]

---

[80] Dr. Charles Till, "Plentiful Energy, the IFR Story, and Related Matters," *The Republic* Jun-Sep 2005.

Lest I be considered a complete solar cynic, let me assure you that I am not. There are some exciting developing technologies that hold great promise, like windows embedded with nanoscale photovoltaics that quietly pump out electricity. But this book is meant to address what is arguably a planetary crisis on several levels. Looking at even the aforementioned cutting-edge system with a projected output of just 50 MW by 2008, it's pretty clear that solar has too far to go to provide a substantial percentage of the world's electricity needs in the foreseeable future. We should welcome with open arms every megawatt solar is able to bring to the party, don't get me wrong, and I think it's worth paying more for clean energy, if necessary. We'll just need a lot more than solar can provide, a lot sooner (and cheaper) than it can conceivably provide it.

In projections of world energy use to the year 2030, the International Energy Agency (IEA) predicts that the category of renewable energy sources that includes "geothermal, solar, wind, tidal and wave energy will grow faster than any other primary energy source, at an average rate of 5.7% per year over the projection period. But [their] share of world demand will still be small in 2030, at 2% compared with 1% in 2002, because they start from a very low base."[81] Even if the IEA is wrong, it can't be *that* wrong.

Wind Power

With those statistics in mind and without wanting to seem dismissive of wind power, there seems to be little point in digressing from my purpose of this chapter, which is to provide a brief look at purported solutions to our planet's energy quandaries and point out unfortunate shortcomings where

---

[81] IEA, "World Energy Outlook 2004," (Paris: International Energy Agency, 2004).

they exist. Wind power, like solar (and, to a lesser extent, biofuels) seems terrific on a visceral level, and like many alternative energy systems it begs to be scaled up to global size. But in reality wind power suffers from serious problems, not the least of which is that the wind is a fickle provider.

When calculating the generating potential of wind, solar, or any other electrical generating system, it's helpful to use their capacity factor. This is simply a ratio of the amount of actual energy they produce in a year (in kilowatt-hours, usually) vs. the amount of energy they would have produced if they ran full bore all year. This is how we calculated the 16% solar efficiency in the preceding section. Many agencies and wind power advocates assume a capacity factor of 40% when estimating the cost of wind-generated electricity, yet the reality seems to be about half that.

Florida Power & Light has the largest amount of installed wind capacity in the United States, with nearly 4,000 MW of generating potential, yet their capacity factor is a dismal 21% overall. On the other side of the country the aggregate capacity factor for California's five largest wind farms is likewise 21%.[82] So why does the National Renewable Energy Lab insist on using a capacity factor of 37% when calculating the costs of generating electricity with the wind? These sites were obviously chosen for their dependable wind potential, so it's inconceivable that the 37% figure is even close to realistic. Calculate the real cost of generation and strip away the 1.8¢/kWh federal wind subsidy and it becomes readily apparent that wind power is hardly as economical as it's touted to be.

Looking to Europe, where wind generation is being pushed more steadily than in the USA, the picture is hardly better. Germany has a wind turbine capacity of 17 gigawatts (17,000 MW) that would provide a notional capac-

---

[82] P. E. Donald E. Lutz, "Wind," in *Truth About Energy* (2007).

ity of 14% of their total energy demand, yet they actually generate only 15% of that amount, a little over 2% of the country's requirements.[83] Denmark's vaunted windmills provide less than 20% of that sparsely populated nation's energy needs, with Sweden and Norway providing back-up when the wind is uncooperative. The other 4/5 of the Danes' electricity is generated by fossil fuels.[84] The wide-spread belief among wind power advocates that Denmark is chugging merrily along on wind power is unfortunately a delusion. On the other hand, when the wind really gets to howling Denmark sometimes ends up with an electric-ity glut and is forced to sell electricity to its neighbors at uneconomic rates.

One of the most optimistic scenarios for massive produc-tion of windpower comes, surprisingly enough, from the U.S. Dept. of Energy. In early 2008 they issued a report suggesting that up to 20% of the USA's electricity needs could be provided by wind power by the year 2030.[85] The same sort of expensive grid upgrades mentioned for solar power would likewise need to be implemented for this sort of major scaling up of wind power, however.

Would it be economically feasible? Let's take a look at a plan by T. Boone Pickens, the famous Texas oil billionaire, to build the world's largest windfarm with a peak capacity of 4 GW. In September of 2007 he told the *Wall Street Journal* that the cost would be $6 billion.[86] By June of 2008 that estimated cost had doubled to a cool $12 billion, $2 billion of which will

---

[83] Rowan Hooper, "Uk Wind Power Takes a Battering," *New Scientist* Nov 12, 2005.

[84] Mattias Akselsson, "The World's Leader in Wind Power," in *Scandi-navica.com* (Sep 2004).

[85] "20% Wind Energy by 2030," (US Dept. of Energy, May 2008).

[86] Mark Gongloff, "Keeping up with T. Boone," *The Wall Street Journal* Sep 19, 2007.

be needed to build the transmission lines to link the system into the grid.[87]

Given the performance of wind farms in prime wind areas of California and Florida, there's every reason to believe that the capacity ratio of Pickens' turbines will be 20% at best, meaning that his 4GW wind farm would translate to an actual average output of about 800MW. Assuming that the ballooning costs don't increase even further by the time it's built, the cost per gigawatt will be a staggering $15 billion.

There is little doubt that wind power will continue to develop, with many countries having systems on their drawing boards. But like solar, multiplying the capacities of systems that provide a mere pittance of the world's current electrical demand, even under best-case scenarios like the DOE projections, will still fall far short of meeting even today's electrical requirements. In a world where energy demand is rising dramatically, the vast majority of our energy needs (both electric & non-electric) will have to be met by other types of systems.

A note about subsidies

Proponents of renewable energy systems, particularly wind and solar, are frequently quick to lament how nuclear power has been unfairly subsidized in comparison to renewables. Statistics, if they're brought into the argument at all, are presented in such a way as to obscure the true economic realities. Often the entire amount of money that the U.S. government has invested in nuclear power research since WWII is used as a measure of this unfair subsidization. This is disingenuous at best. Of course the government funded nuclear power research — for security reasons, if nothing else. Much of the research was double-pronged, involving both military

---

[87] Bruce Gellerman, "Don't Mess with Texas Wind," in *Living on Earth* (USA: Jun 6, 2008).

and civilian uses (eg. nuclear-powered naval vessels whose reactor principles could apply to civilian generators).

On the other hand, renewable energy systems are treated to massive subsidization today. Ethanol production from corn is subsidized by you, the taxpayer, to the tune of over fifty cents a gallon! Solar and wind power are both heavily subsidized as well. Nuclear power, on the other hand, while producing nearly 20% of the electricity in the USA, is perking along without any subsidies (though of course originally there were tax breaks involved with the building of the plants). Comparing the subsidies of these technologies with the amount of electricity each is producing puts nuclear way down on the subsidies/kWh list. In point of fact, if the nuclear power industry could avoid the interminable delays and holding actions of antinuclear activism and instead build and operate plants in an expeditious manner, nuclear power advocates would love to compete on an unsubsidized level playing field with renewables.

Hydroelectric Power

Back in the heady days before environmental impact statements, when government could dictate progress come hell or high water, hydroelectric dams sprouted like a beaver's utopia. But the rise of the environmental movement and a greater appreciation for the social costs of relocating residents of soon-to-be-submerged towns have brought dam building to a screeching halt in many countries. China and India, desperately hungry for power, have ambitious hydropower projects, most famously (or infamously) the giant Three Gorges Dam complex on the Yangtze River. It will be the largest hydroelectric dam in the world, five times larger than the Hoover Dam. The power it produces will be about 25% more than the current largest dam complex in Canada. China, already boasting hydropower capacity almost double that of the United States (which is second in capacity), has at least nine major dam building projects in the works.

China also has the dubious distinction of having experienced the most disastrous dam failure in history. When Typhoon Nina hit China in 1975, the failure or intentional "preventative" destruction of some 62 dams occurred in an almost unbelievable domino effect. By the time the full impact of the disaster was added up some 171,000 people were dead (from direct and indirect causes) and millions were left homeless. While not the first disastrous dam failure, it dwarfed any others before or since.

Hydroelectric power comprises the vast majority of electricity production from renewable resources today, dwarfing the output of wind and solar projects. But given the power of environmental groups in most industrialized countries today, it is doubtful that hydro's share of total power capacity will be increasing there in the future. Wind and solar developments have already run into the NIMBY factor (Not In My Back Yard), but hydro is even tougher because it collides with the NOMBY problem (Not Over My Back Yard—and front yard, for that matter!). There is also the disruption of fish populations to consider, especially in watersheds where salmon spawn.

The reticence to exploit hydroelectric power does not necessarily apply to developing countries, however. The prodigious hydroelectric potential of the Congo River has remained relatively untapped due to the ongoing violence in that region that's claimed over three million lives. Recent stability, however, has resurrected plans for the Grand Inga Project, a hydroelectric generation scheme that dwarfs even China's Three Gorges system. It relies on a natural drop of the Congo River of some 100 meters, and thus would require only a modest dam and reservoir relative to its enormous generating capacity. It is estimated that this hydro complex would produce some 39 gigawatts, enough to supply the full amount of electricity used by the entire continent today. Of course per capita use of electricity in Africa is extremely low and will inevitably increase substantially in the future, yet bringing a hydro project like this one online would be a huge boost to Africa's infrastructure.

Proponents of distributed generation point out that there are plenty of streams and rivers that would easily support micro-hydro projects, small turbines or water wheels for small-scale electrical generation. While these may well be a boon to people who are in a position to take advantage of nearby flowing water sources of sufficient capacity, clearly this is and will continue to be a niche producer, and simply cannot be expected to make a substantive dent in world energy supplies.

As for the giant projects being built in countries where quality control is often a problem, let's just hope for the sake of those downstream that typhoons like Nina's big sister never show up. Hopefully if the Grand Inga Project is undertaken there will be sufficient oversight to prevent shoddy workmanship. The fact that it would utilize a relatively small dam, however, should minimize the potential for future disaster even in a worst-case scenario.

Geothermal Energy

The idea of extracting heat from the earth would seem to be a no-brainer, especially given the fact that oil drilling to the depths necessary to reach very hot rock has become routine. Converting that heat into electricity is an established technology that was demonstrated over a century ago, yet today less than ½ of 1% of the world's primary energy supply is derived from geothermal sources.[88] Iceland has taken advantage of its extremely obvious geothermal gifts, and there are some substantial generating plants located in California and elsewhere.

---

[88] IEA, "Renewables in Global Energy Supply, an IEA Fact Sheet," (International Energy Agency, Jan 2007).

Proponents of geothermal sometimes cite statistics to demonstrate that all the energy required by humans can be supplied by geothermal many times over. But then again, proponents of nearly every purported panacea cite such statistics, such as the claim that there is enough wind between the Rockies and the Mississippi River to supply all the energy needs of the United States. Just because raw energy is there doesn't mean that it's necessarily economical or even possible to corral it.

That being said, it would seem perhaps more logical than most such arguments that the heat of the earth's mantle could be tapped quite readily. Germany has invested heavily in geothermal R&D, committed as they are to eschewing nuclear and coal power. Thus there was more than a ripple of concern when Basel, Switzerland was shaken by an earthquake and several aftershocks attributed to a geothermal drilling project there.[89] The quake, while measuring 3.4 on the Richter scale, wasn't sufficiently powerful to cause major damage, but the public relations damage was another story. The project was immediately shut down, yet tremors — some of them substantial — continued to be felt for weeks. The quakes were felt in Germany, too, which paid more than a little attention given their own efforts in this field.

The Swiss were using what is sometimes referred to as the Hot Dry Rock or Hot Fractured Rock system. It consists of drilling two deep holes some distance apart to sufficiently hot layers, hydraulically fracturing the rock between them, then pumping water into one and retrieving steam from the other one that has been generated by the water seeping through the hot rock between the two. Research into such systems has been going on for over thirty years, and much has been learned.[90]

---

[89] Christine Lepisto, "Geothermal Power Plant Triggers Earthquake in Switzerland," in *Treehugger.com* (Jan 21, 2007).

[90] Idaho National Laboratory, "The Future of Geothermal Energy," (Renewable Energy & Power Dept. of INL, 2006).

While induced seismic activity due to such projects comes as no surprise to those involved in the field, an MIT-led study cast doubt on the idea that such quakes could be large enough to inflict significant damage.[91]

Among the many difficulties encountered in such geothermal projects is the tendency of the injected water to establish channels between the two boreholes, which quickly cools the rock surrounding them and diminishes the transferred heat. Even without such localized cooling, the gradual cooling of the greater area can eventually occur, necessitating either a shut down of the project or a respite while the earth is allowed to heat up sufficiently to resume steam production. Working up to ten kilometers deep underground, this is not a particularly easy technology to deal with. Yet the vast amounts of heat literally beneath our feet is a tempting target for all of us who desire clean and abundant energy.

It is certainly possible that great strides will be made in geothermal heat production and that eventually we may see it producing vast amounts of electricity. Unfortunately that day is not yet here despite decades of research. Whereas future R&D efforts in this field are clearly warranted, it isn't yet ready to assume a dominant role in supplying humanity's energy needs, nor can we predict when or if that day will arrive. If we wish to eliminate or at least drastically reduce greenhouse gas emissions in the near future, we'll have to look toward technologies more readily at hand, while continuing to encourage further development in this promising area.

Hydrogen

If you asked a random sampling of people on the street to identify an energy panacea for the future, most would probably suggest hydrogen. There has been so much hype and

---

[91] Ibid.

such enthusiastic greenwashing about hydrogen power that one could be forgiven for not noticing that hydrogen isn't actually an energy source at all, but merely an energy carrier.

Hydrogen is an important trace constituent (0.5 parts per million by volume) of the atmosphere, but it does not exist in its elemental state on earth to be collected and used as we would wish. It must be split off of compounds in order to be segregated for use as a fuel. Anyone who's taken a high school chemistry class is probably familiar with electrolysis being used to split water into its components of hydrogen and oxygen. This technique is most commonly envisioned as the way we could liberate hydrogen for use as a fuel for our vehicle fleets around the world, though in reality the hydrogen used today is mostly derived from natural gas, producing large amounts of global warming gases in the bargain. The truly starry-eyed envision solar and wind farms generating enough power to not only supply our normal electrical needs, but enough excess power to produce hydrogen for both transportation and electricity for those times when it's dark and the wind isn't howling. As we saw earlier, it would be grossly understating the case to call that unrealistic.

Meanwhile everyone from President Bush to British Petroleum has jumped on the (still parked) hydrogen bandwagon. Why not? The technologies are still so immature, the costs so stratospheric, and the technical problems of storage and distribution so daunting that it makes the perfect greenwash. Couple that with the enticing cliché of driving down the street producing naught but clear water as exhaust and the scenario is almost irresistible, especially to those ignorant of the technologies, which includes nearly everyone — including President Bush, I'm sure. But as evidence of the immediacy of global warming consequences grows year by year, the utopian "hydrogen economy" seems to just keep receding farther and farther in the distance.

Joseph Romm worked in the U.S. Department of Energy (DOE) under President Clinton, overseeing research and development of clean energy systems. He has written a comprehensive and insightful book called *The Hype About Hydrogen* which is a must-read for anyone who wishes to have an unvarnished view of the vaunted "hydrogen economy." Salon.com describes him as concerned "that the hyperbolic promotion of hydrogen fuel-cell cars as the answer to our energy woes is a scientific and technological wild goose chase, engaged in at our peril while the global-warming clock rapidly runs down."[92] In keeping with my intention of brevity I will enthusiastically recommend Romm's book and mention only a couple of the more salient points here.

Hydrogen is notoriously difficult to contain. It wants to get out of anywhere you stuff it, and the idea of producing it on a scale vast enough to power the world's transport seems almost like a small detail (it's not!) when compared to the prospect of storage and distribution.

> With the current generation (circa 1990) of highly insulated double-walled vacuum-jacketed storage tanks manufactured by the German Aerospace Research Establishment DFVLR (Stuttgart), the liquid hydrogen will evaporate at a rate of about 8% per day. Because the evaporation increases the pressure on the tank wall, the gaseous hydrogen must be vented to the atmosphere to keep the tank from rupturing. Investigators at Los Alamos National Laboratory found that in a 1979 liquid hydrogen-fueled Buick they were testing, a full tank of liquid hydrogen would evaporate in about 10 days.[93]

---

[92] Salon.com Interview w/Joseph Romm, "Just Say No, to Hydrogen," in *Salon.com* (Apr 29, 2004).

[93] Harry W. Braun, *The Phoenix Project: Shifting from Oil to Hydrogen* (Sustainable Partners Inc, Dec 1, 2000).

The reader will be forgiven if contemplating the result of venting a tankful of hydrogen into a closed garage brings to mind images of the Hindenburg.

Lest my cynicism about the benign nature of oil companies be revealed, I must nevertheless raise the possibility that one reason they're involved at all in the R&D for the hydrogen utopia is that they'd love to be the ones in charge of its vastly complicated and absurdly expensive infrastructure. It doesn't hurt that the majority of R&D funding is going straight into their pockets even now, of course. If something is going to take the place of gasoline, Big Oil certainly intends to be holding the controls.

They're some mighty expensive controls, too. Romm cites a study by Argonne National Laboratory indicating that "with current technologies, the hydrogen delivery infrastructure to serve 40% of the light duty fleet is likely to cost over $500 billion."[94] This starkly exposes the critical chicken and egg problem that can hamstring any new technology. Who will pay for an energy infrastructure of such prodigious cost with no guarantee that technological developments in the near future won't make it obsolete? It's a certainty that nobody will buy hydrogen cars until the distribution and fueling systems are in place. And what possible motive would oil companies have to hang a trillion dollars out on the line in the hopes of replacing a system that they already control and which brings untold wealth into their coffers like clockwork?

The American Physical Society is an esteemed organization of physicists whose purpose is to "advance and diffuse the knowledge of physics." They publish the world's most prestigious and widely read physics research journals, and in the course of their work they examine a great variety of concepts, including hydrogen power. Here is their commentary on Bush's

---

[94] Folga Mintz, Molburg, Gillette, "Cost of Some Hydrogen Fuel Infrastructure Options," (Transportation Technology R&D Center, Argonne National Laboratory, Jan 16, 2002).

"Hydrogen Initiative" in a nutshell (an appropriate receptacle) following a study of its possibilities:

> **Major scientific breakthroughs required for the Hydrogen Initiative to succeed, panel finds.**
>
> The American Physical Society's Panel on Public Affairs (POPA) today released a report that analyzes the Hydrogen Initiative. President Bush proposed the initiative in his 2003 State of the Union Address. The Hydrogen Initiative envisions the competitive use of hydrogen fuel and a hydrogen-fueled car by the year 2020.
>
> The POPA report concluded that major scientific breakthroughs are required for the Initiative to succeed. The most promising hydrogen-engine technologies require factors of 10 to 100 improvements in cost or performance in order to be competitive. Current production methods are four times more expensive than gasoline. And, no material exists to construct a hydrogen fuel tank that meets the consumer benchmarks. A new material must be discovered.
>
> These are very large performance gaps. Incremental improvements to existing technologies are not sufficient to close all the gaps. Significant scientific breakthroughs are needed. According to Peter Eisenberger, chairman of the committee that drafted the report, "Hydrogen storage is a potential show stopper."[95]

Yeah, it sounds pretty grim. But here's the real deal breaker: A group of Cal Tech/JPL scientists who were noodling over the possibilities of a hydrogen economy got to wondering

---

[95] APS Panel on Public Affairs Energy Subcommittee, "Hydrogen Initiative Report from American Physical Society Panel Released," (American Physical Society, Mar 1, 2004).

what would happen with all that free hydrogen that would most certainly be leaking into the atmosphere if the world's vehicles were converted. Even if one could somehow have a technological breakthrough to store the hydrogen in solid form, the liberation into a form usable in fuel cells would most certainly provide a chance for plenty of hydrogen to escape. Think about gas leaks in cars, then think about hundreds of millions of cars (including — especially — those in low tech societies). Hydrogen is harder to contain than gasoline by far. Even in the best-case scenarios, and assuming technological leaps in storage technologies that at present are only dreamed of, there'll still be an awful lot of hydrogen being liberated into the atmosphere.

The Cal Tech study[96] came to a rather startling conclusion: If the world's vehicle fleet were converted to hydrogen as a fuel, the resulting leakage would very possibly cause the levels of hydrogen in the stratosphere to increase to the point that the ozone layer would be seriously damaged. Just when we started getting a handle on the ozone layer problem by outlawing CFCs (and the ozone hole problem is hardly solved yet), along come the hydrogen true believers ready to make the CFC problem look like a mere warm-up. Is the Cal Tech study definitive? Not necessarily. The mechanism by which the hydrogen would break down the ozone layer is pretty well understood, but the process of absorption of atmospheric hydrogen by soil bacteria is still quite hazy. Could the earth's hydrogen-gobbling mini-denizens absorb the leakage of a billion cars to keep the stratosphere free of excess hydrogen? Nobody knows.

Are you willing to bet the ozone layer on it?

---

[96] *Potential Environmental Impact of a Hydrogen Economy on the Stratosphere,* Tracey K. Tromp,1 Run-Lie Shia,1 Mark Allen,2 John M. Eiler,1 Y. L. Yung1, California Institute of Technology (2003)

If hydrogen is the most promising solution to our energy dilemmas, then we're in a world o' hurt. As I said in the beginning of this section, it isn't actually an energy source anyway. Obtaining it from natural gas or coal or other fossil fuels makes no sense, neither economically nor environmentally. And despite the research being done today on storage technologies, the threat of damage to the ozone layer is a distressingly real possibility. Even if we could obtain all the hydrogen we wanted from clean solar power (which we can't), there is a very high likelihood that we wouldn't be able to use it anyway for fear of rendering our planet unlivable.

Fusion

The promise of commercial power generation using fusion reactors is the holy grail of energy production. When you describe it the process sounds relatively simple: Two light nuclei are brought together with sufficient energy to overcome the electrostatic force between them and fuse together (hence "fusion") to form a heavier nucleus and, in the process, release energy. The "sufficient energy" part is the kicker, though. In the fusion reactors envisioned as the first generation, plasma containing deuterium and tritium (isotopes of hydrogen) would be heated to temperatures about ten times hotter than the sun's core in order to induce the fusion reaction.

As impossible as that may sound, it has been done for about a second, though the energy produced was less than the energy put into the system. The promise of fusion's unlimited potential, though, has led to an international effort to push the technology forward to commercial applicability. The ITER project (International Thermonuclear Experimental Reactor) began in 1985 as a collaboration between the European Union (through EURATOM), the USA, the then Soviet Union and Japan. The stated purpose was to "demonstrate the scientific and technological feasibility of fusion energy for peaceful pur-

poses." The USA has been a fickle participant, though, drop-
ping out of the consortium and thus threatening its viability,
then re-engaging in 2003 as the politics of global warming
have made the search for solutions more urgent. Or is it because having two oilmen at the top of the U.S. government has
made greenwashing more urgent?[97]

Politics is a big part of ITER. Opposition to the ITER project has been generated most vociferously from the same environmental groups that oppose nuclear fission power. "Pursuing
nuclear fusion and the ITER project is madness," said Bridget
Woodman of Greenpeace. "Nuclear fusion has all the problems
of nuclear power, including producing nuclear waste and the
risks of a nuclear accident."[98]

Whoa, lady! Take a chill pill. I'm sorry, but I have little patience for hysteria when it comes to discussing the serious problems that face us today. While I consider myself a very serious
environmentalist (judge for yourself after reading this book),
off-the-wall statements like that are simply out of bounds. This
is a typical knee-jerk reaction to anything with the word "nuclear" in it, and is either based on the rankest ignorance (in
which case she should hardly be speaking for her organization)
or an appalling disingenuousness (ditto).

Fusion reactors, should they be proven viable as they almost surely will be — eventually — would produce a pittance
of nuclear waste with so short a half-life that it would be harmless within ten to a hundred years. The accident risk is likewise
overblown, since it would be impossible for a fusion reactor to
undergo a runaway chain reaction. It's not the safety that's the
problem; it's the time it will take to make the concept commercially viable.

---

[97] I realize I'm stretching credibility to call George Bush Jr. an oilman.
[98] Dallas Kachan, "$13b Nuclear Fusion Research Agreement Signed," in
Cleantech.com (Nov 21, 2006).

There's an old joke among nuclear physicists that says practical fusion is only about forty years away…and always will be.
Yeah, I know, those physics jokes are real knee-slappers, aren't
they? Did you hear the one about Werner Heisenberg getting
pulled over for speeding?[99] But I digress. Estimates of fusion
reactor deployment from the physicists and engineers most
knowledgeable about the subject range from about forty to a
hundred years. Maybe we'll be surprised and they'll be able
to do it sooner. But unfortunately our planet doesn't seem in-
clined to give us the time. While continuing research into fusion power makes sense both from a pure research standpoint
and as a long-term solution to provide earth's inhabitants with
clean, safe, and unlimited power in the future, we'd better do
something serious with the technologies available to us today.

Nuclear Controversy

Antinuclear activists (hereinafter to be referred to as anties
in the interest of brevity) have long warned of the legacy of
nuclear waste that we're leaving for untold future generations. Many have gone farther to decry the discharge of radioactive materials from power plants. Since there has been
so much controversy over the myths and realities, here are
some statistics from a study done at Oak Ridge National
Laboratory about power plant discharges in 1982:[100]

- A typical power plant annually releases 5.2 tons of uranium
  (containing 74 pounds of fissile U-235, used in both power
  plants and bombs) and 12.8 tons of thorium.

---

[99] "Do you know how fast you were going?" asks the cop. "No," says Werner, "but I can tell you exactly where I am."
[100] Oak Ridge National Laboratory Alex Gabbard, "Coal Combustion:
Nuclear Resource or Danger?," (Feb 5, 2008).

- Total U.S. releases for 1982 came to 801 tons of uranium (containing 11,371 pounds of U-235) and 1971 tons of thorium.
- Worldwide releases totaled 3,640 tons of uranium (containing 51,700 pounds of U-235) and 8,960 tons of thorium.

Considering the longevity of radioactive materials in the environment, the study also looked at the cumulative releases and came up with some sobering projections. By the year 2040, cumulative releases of radioactive materials from these power plants will have reached the following levels:

- U.S. releases: 145,230 tons of uranium (including 1,031 tons of U-235) and 357,491 tons of thorium
- World releases: 828,632 tons of uranium (including 5,883 tons of U-235) and over two million tons of thorium.
- "Daughter products" produced by the decay of these isotopes include radium, radon, polonium, bismuth, and lead.

Why is this not splashed all over the front pages? Who in their right mind can consider this acceptable? Shouldn't these numbers alone, published by one of the USA's most respected national laboratories, spell the immediate demise of the nuclear power industry?

Well, let's not get out the torches and pitchforks just yet for a trip down to the closest nuclear power plant, because while these figures aren't in dispute, they are not referring to nuclear power plants at all. These are the radioactive release figures for coal-fired power plants!

Population exposure to radiation from coal-burning power plants is over a hundred times higher than anything conceivably coming out of nuclear power plants. And while a portion of these isotopes is spewed out of the power plant's smokestacks, the rest are concentrated in the coal ash, which is then summarily dumped.

> Large quantities of uranium and thorium and other radioactive species in coal ash are not being treated as radioactive waste. These products emit low-level radiation, but because of regulatory differences, coal-fired power plants are allowed to release quantities of radioactive material that would provoke enormous public outcry if such amounts were released from nuclear facilities. Nuclear waste products from coal combustion are allowed to be dispersed throughout the biosphere in an unregulated manner. Collected nuclear wastes that accumulate on electric utility sites are not protected from weathering, thus exposing people to increasing quantities of radioactive isotopes through air and water movement and the food chain.[101]

If this isn't crazy enough for you, ponder this little factoid: The energy content of the nuclear materials released into the environment in the course of coal combustion is greater than the energy of the coal that is being consumed. In other words, coal consumption actually wastes more energy than it produces, and contaminates the environment with radioactive materials over a hundred times more than nuclear power plants.

The hysteria of those who vilify nuclear power looks a hundred times more irrational when these facts are known. It is difficult to resist the temptation (so I won't) to compare antinuclear fanaticism with religious extremism. Both have their high priests (or priestesses, as in the case of Helen Caldicott, the doyenne of antinuclear hysterics). Both have legions of followers who

---

[101] Ibid.

don't really understand the mysteries of the subject at hand but instead place their trust in their respective priesthoods and then passionately espouse whatever they're told. Both pour massive energy and money into the coffers of their organizations, which do their best to influence legislation. And both have their true believers embedded in the government, basing decisions that affect all of us on emotional appeals and a repudiation of logic and rationalism. It's difficult to find any parallel in history for an ideology being constructed around a physical process without hearkening all the way back to the seventh century B.C.E. in Greece, when Prometheus was venerated for bringing fire to humankind. Only this time the bringers of fire are being vilified.

"So, we've got to know that there is a conspiracy out there and the conspiracy is against the people," rants Helen Caldicott.[102] Let's examine this allegation with a bit of logic. There are many thousands of nuclear physicists and engineers who are more than willing, nay eager, to support the use of nuclear power. Few would argue that these people, who are generally a cut above the hoi polloi in the smarts department, are Strangelovian monsters who care nothing for their children and grandchildren as they push an agenda of poisoning the world for their descendents. Nor would any substantial number of those very smart people have made financial decisions to invest in the nuclear power industry, since it's been on the skids now for at least a few decades.

Conspiracies of this magnitude, involving tens of thousands of scientists, engineers, accountants, technicians, and politicians[103] are all the more ludicrous when you consider that all these people would knowingly and maliciously be dooming their own progeny to lives of misery and untimely death. Yet

---

[102] Helen Caldicott speaking at Real Goods Alternative Energy Store, Hopland, CA, June 26, 1999

[103] Caldicott accused Jimmy Carter in the aforementioned Hopland speech of covering up the data about Three Mile Island.

such absurd charges are tossed about repeatedly, despite the complete lack of any rationale for anyone to so clearly work against their own well-being and that of their families.

The allegation that nuclear plants are routinely emitting radiation and that it's dangerous to live near them is a frequent charge of nuclear opponents. In a 2005 interview Caldicott claimed, "The literature is replete with malignancy in people who live near reactors. But because of the latent period of carcino-genesis, the incubation time for cancer is five to six years. You have to wait for a while and do a decent epidemiological study to assess what's going on." Helen must have missed this one:

> In 1991, the National Cancer Institute in the U.S. conducted what might be considered a "decent epidemiological study" of deaths from 16 types of cancer, including leukemia, in 107 U.S. counties "containing or closely adjacent to 62 nuclear facilities," all of which had been built before 1982. The survey compared cancer death rates before and after the facilities went online with similar data in 292 counties without nuclear facilities. After four years of research, the team of epidemiologists found no general increased risk of death from cancer near nuclear facilities. In some counties, the relative risk for childhood leukemia from birth through 9 years dropped a statistically insignificant few hundredths of a point after the startup of a local nuclear facility. The areas surrounding four facilities, including San Onofre, showed significantly lower rates for leukemia in teenagers compared with the rest of the country. A University of Pittsburgh study of the area within a five-mile radius of Three Mile Island showed no statistically significant increase in cancer rates 20 years after the accident at the reactor in 1979. What's more, neither soil nor air samples in the area around Three Mile Island have been kept from the public [contrary to Caldicott's claims].

According to the Carter-era EPA, close to 10 percent of some 800 milk samples from local dairy farms the month after the accident showed trace amounts of radioactive contamination. But the highest concentration was still 40 times less than what showed up in milk after the fallout from Chinese nuclear testing in October 1976 that passed across the United States.[104]

Another oft-cited study condemning nuclear reactors is the so-called "Tooth Fairy Study" which attempted to link nuclear power plants with supposed releases of strontium-90 by studying teeth of children downwind of nuclear power plants. Eight states (Connecticut, Florida, Illinois, New Jersey, New York, Pennsylvania, Minnesota and Michigan) undertook an examination of this study out of understandable concern for their citizens. Every one of them found the study to be without merit. Here's what the New Jersey Commission on Radiation Protection reported to their governor in 2004:

> The Commission is of the opinion that "Radioactive Strontium-90 in Baby Teeth of New Jersey Children and the Link with Cancer: A Special Report," is a flawed report, with substantial errors in methodology and invalid statistics. As a result, any information gathered through this project would not stand up to the scrutiny of the scientific community. There is also no evidence to support the allegation that the State of New Jersey has a problem with the release of Sr-90 into the environment from nuclear generating plants: more than 30 years of environmental monitoring data refute this.[105]

---

[104] Judith Lewis, "Green to the Core? — Part 1," in *L.A. Weekly* (Nov 10, 2005).

[105] Eric McErlain, "Real Science Refutes The "Tooth Fairy"," in *NEI Nuclear Notes* (Mar 4, 2005).

It is especially easy to dupe a gullible populace if a question is complex, when most people would never dream of trying to understand the realities of the subject at hand. So real facts can be mustered to support an ideological position despite the most blatant deception being intended. These tactics are used repeatedly by anties. One doctor of health physics at Oak Ridge National Laboratory, outraged by such deceptions, formulated his own analogous argument to show how it's done:

> "Potassium-40 is a lethal toxin, which is blithely distributed to the public by grocery store chains all over America every day of the year. This deadly isotope remains radioactive for nearly 13 billion years, and emits extremely high-energy gamma rays (1.5 million electron volts) and beta particles (1.3 million electron volts). This material is present in high concentrations in many of the foods that these stores foist on the public, with full knowledge of the US government. Radiations of this type and energy are well known for causing leukemia, breast cancer, and fatal and non-fatal birth defects."
>
> Everything said here is true—K-40 is a naturally occurring radioisotope of potassium—many foods like bananas, fruit flavored sherbets, and potassium salts contain lots of K-40. "People" contain lots of K-40. It has a very long half-life, about 1.3 billion years, and does emit high-energy radiations. Radiation has been linked to cancer. Are the grocery stores thus in collusion to irradiate and cause cancer in the American public? No. But I could make it sound that way if I wanted to. Unfortunately, many of the arguments (pro-nuke, anti-nuke, pro-life, pro-choice) carried in the media these days

are dominated by people who have made a living out of dealing in hyperbole instead of the truth.[106]

Notice how billions of years are invoked, calling to mind the vast ages of nuclear waste longevity that virtually everyone has heard. This is a frequent ploy by those who wish to scare and mislead the public about nuclear matters, as in this tidbit from a Sierra Club publication:

> Unfortunately, there is no real way to be rid of radioactive materials, because some fission byproducts and nuclear wastes remain hazardous for extraordinarily long periods of time. For example, it takes 4.5 billion years for just half of the atoms in Uranium-238, the primary source of nuclear fuel, to disintegrate.[107]

The Sierra Club is absolutely right here, there is no way to be rid of U-238, since there is plenty of it widely distributed in the earth's crust, and in sea water as well. Is that a bad thing? With a half-life of 4.5 billion years, it sounds like a scary toxin that we should be really worried about. Or at least that's what it seems they would have you believe. If you don't understand the concept of the half-life of radioactive materials — which the majority of the public does not — then this sounds like a horrible situation.

Without putting too fine a point on it, the shorter the half-life the more immediately dangerous the element, and generally vice versa. The type of radiation also makes a difference. U-238, with a 4.5 billion year half-life, is hardly dangerous. It is frequently used as ballast in both sailboats and aircraft, and

---

[106] Mike Stabin Ph.D. CHP, "Nuclear and Radiation Safety Issues," *L.A. Times* 2006.

[107] Judith Johnsrud, "Why Nuclear Power Is Not a Solution," *The Sylvanian* Nov 2005-Jan 2006.

has even been used on occasion as a very effective door stop, since it's about 50% heavier than lead. The public knows it more commonly as DU, or depleted uranium, which is used in armor plating and projectiles in modern warfare. One would not want to inhale atomized U-238, but that admonition would apply to quite a few substances. The only reason it can be used as nuclear fuel is because when bombarded with neutrons from U-235 or other fissile elements it can capture neutrons and be transmuted into a different element, which is what actually constitutes the fuel. U-238 could be considered, in this light, as a fuel precursor, what is called in the trade a fertile material.

Categorical thinking clearly can help us make sense of the world. It's an evolutionary advantage to be able to quickly make decisions based on past experience. But sometimes such mental sorting fails us, and sometimes it's positively detrimental to our well-being. We have to be able to recognize when that is happening.

Fear can be one of the most powerful motivators for categorical thinking. Many people take an absolutist stance against nuclear power because they associate it with nuclear bombs, despite the fact that a nuclear power plant in even the worst of worst-case scenarios can't cause a nuclear blast. Antie ideologues will frequently use this confusion and lack of knowledge to conflate the two and reinforce the categorical imperative. After decades of this sort of fear mongering, for many people anything with the word "nuclear" in it is bad. The programming has been very effective.

Politicians use categorical thinking all the time when they perceive it to be to their advantage. A great deal of political pronouncements are designed specifically to trigger visceral responses and knee-jerk reactions in voters. They know that people either don't have or, more often, won't take the time to explore the nuances of policy. There's just too much information out there. Even the politicians themselves, who are making tremendously consequential decisions for the rest of us, resort to categorical thinking all the time. And they often use it as a political weapon even if they know better. In a 1992 debate in New Hampshire against his principle contenders for the nomination, Bill Clinton saw fit to attack a rival with an accusatory, "You're pro-nuclear!" as though he couldn't believe anyone could be that foolish. He knew it would be a crowd pleaser, and he was right.

Today we see some quite unexpected people coming out in favor of nuclear power, very strongly advocating a wholesale switch to it. Stewart Brand, of Whole Earth Catalog legend, is joined in his apostasy by Patrick Moore, one of the founders of Greenpeace. These people have broken out of old categorical thinking modes and are urging others to do so too. For when we're faced with a serious crisis — as they and most of the scientific community believe we are — categorization cannot be allowed to cloud our thinking when considering our options.

The unfortunate tendency to equate the word nuclear with danger and corporate skullduggery has been nurtured by antinuclear organizations for decades, with considerable success. The aforementioned Mr. Moore had the dubious privilege of participating in a panel discussion at the Society for Environmental Journalists' annual conference on October 27, 2006. The Nuclear Information & Resource Service, which inexplicably keeps forgetting to put the "Dis" in front of the second word in their title, linked to a video of the event[108] from their website, where we could watch Mr. Moore fend off two antinuclear spokesmen. The heading on the page hinted at the objectivity of the "discussion," being labeled "Dirty Power — False Promises: Nuclear Power & Climate Change." Meanwhile, PowerPoint-style slides describing the points being raised were flashed onto the screen in accompaniment to the presentation of each of the antie representatives.

---

[108] Nuclear Information & Resource Service, "Think Nuclear Power Can Save the Climate?," (NIRS, Oct 27, 2006).

But when Mr. Moore was presenting his arguments, slides attempting to debunk the points he was making were flashed on the screen in place of the slides he was presenting at the event.

Some of the slides that were intended to tear him down contained footnotes, which was truly bizarre since the lies that they were stating were actually, in some cases, directly refuted by the footnoted source. There was the old square miles vs. miles square solar power flim-flam described earlier in this chapter. That one was off by a factor of 100. Then, in an argument for distributed generation, one slide made the point that with transmission grids which are increasingly strained and inefficient, "By the time electricity reaches the customer, nearly two-thirds of the energy has been wasted through transmission." That the Nuclear Disinformation Service can even trot out such an unbelievable figure is a measure of just how brainwashed they believe their fans to be. But they are as foolish as they are mendacious, for they provided a handy footnote. A quick glance at the citation revealed that the source article[109] was talking about the inefficiencies of antiquated turbine designs, not line loss. A moment's further investigation easily turned up the actual figure for electricity lost in transmission and distribution: 7.2%, of which 40% is due to transformers.[110]

There were other citations of long-debunked studies and outright lies, but I won't belabor the point. Such an approach may succeed with the true believers, but for anyone with a little knowledge and a healthy dose of skepticism it casts doubt on the valid points that were made elsewhere in the presentation, and there were some that bear considering. Part of the discussion dealt with safety and oversight, expressing valid concerns about the efficacy of the Nuclear Regulatory Commission and

---

[109] Richard Munson, "Yes, in My Backyard: Distributed Electric Power," in *Issues in Science & Technology* (Winter 2006).

[110] "Technology Options for the near and Long Term " (U.S. Climate Change Technology Program, Aug 2005).

some of their lapses. It's a shame, really, that deceit and disinformation are so blithely and frequently employed by those who fault their enemies for those very tactics.

Few would dispute that nuclear power can and should be made even safer than it has been in the past. But as it stands today, nuclear has a stellar safety record. With the unfortunate exception of Chernobyl, which resulted directly in 56 deaths and in deleterious consequences to many others, the nuclear power industry has been far more benign than any other type of power generation. Even adding Chernobyl into the mix (with its faulty plant design that was only used in Russia and is no longer employed), far more people have been injured and killed due to hydropower, the oil industry, and even natural gas. Not a single person has ever been killed due to a radiation accident in the entire history of the U.S. commercial nuclear power industry. Yet the very week I wrote this paragraph over a hundred coal miners died in a mining accident in Russia. Granted, solar and wind have a fairly harmless record so far, though a lot of birds (and even more bats) would eagerly take issue with me on that point.

Which brings us back to coal. A coalition of national environmental groups called *Clear the Air* commissioned a study from Abt Associates, one of the largest government and business research and consulting firms in the world. This firm has provided the Environmental Protection Agency (EPA) and the Bush administration with analysis of many of the agency's air quality programs. Knowing the track record of Bush's EPA and its antipathy to alternative energy, one might reasonably suspect that this firm's conclusions would hardly be slanted on the side of environmentalists. Thus their conclusion may surprise you: Some 24,000 people die prematurely in the United States each year just from the effects of soot from coal-fired power plants, by an average of 14 years. The study also pegged the annual total health costs associated with soot from power plants at over 167 *billion* dollars![111]

We're just talking about soot here, not the acid rain, heavy metals and other nasty materials scattered through our environment by both smokestack emissions and solid ash disposal. Nor are we even considering the effects of the staggering carbon dioxide emissions that are the main contributor to global warming.

So what would it take to get disingenuous demagogues to quit harping about Three Mile Island? Its monolithic concrete building with the rounded top is called a containment building, as seen at nuclear plants around the world. The reason they call it that is because it's meant to contain radioactive material in the event of an accident. Chernobyl didn't have one. TMI did, and it did its job. The only radiation released at TMI was a purposeful venting of some readily dispersed gases, and that was on hindsight considered to have been a controversial (in terms of P.R.) and perhaps unnecessary precaution. Antie groups tried their best to allege harm due to this most celebrated of U.S. nuclear accidents, but were unable to come up with anything that could stand the scrutiny of science and the law. The area around TMI was sampled for every possible sort of radioactivity more than any single patch of ground in history, yet all that anyone was able to come up with were unsupported allegations of nonspecific harm.

Within weeks of the accident at Three Mile Island, attorneys filed a class action suit encompassing over 2,000 personal injury claims. The suit dragged on for nearly twenty years. The conclusion of the judge, who had given every benefit of the doubt to the plaintiffs, was clear:

> The parties to the instant action have had nearly two
> decades to muster evidence in support of their respective

---

[111] J.R. Pegg, "Coal Power Soot Kills 24,000 Americans Annually," in *Environment New Service* (Jun 10, 2004).

cases. As is clear from the preceding discussion, the discrepancies between Defendants' proffer of evidence and that put forth by Plaintiffs in both volume and complexity are vast. The paucity of proof alleged in support of Plaintiffs' case is manifest. The court has searched the record for any and all evidence which construed in a light most favorable to Plaintiffs creates a genuine issue of material fact warranting submission of their claims to a jury. This effort has been in vain.[112]

So if you hear anybody arguing against nuclear power based on the legend of Three Mile Island, please tell them about real radioactive discharges and where they could more effectively channel their outrage, if they feel compelled to do so. Point them in the direction of a coal-fired power plant.

Nuclear Power — Fission Style

The year 1979 saw the publication of a book proposing a controversial concept called the Gaia hypothesis. In essence, it conceives of the earth as a self-regulating and self-correcting super organism, maintaining through its complex interrelated environmental mechanisms the conditions most conducive to life. While criticized by some as being teleological, the concept was embraced by many environmentalists the world over and made its author, the British scientist James Lovelock, a virtual icon of the environmental movement.

Yet the advance of global warming seems to indicate that any such overarching mechanism that may be in place to maintain a biologically friendly homeostasis shows signs of being overtaxed to the point of failure. Lovelock, now 88, has embraced what many of his erstwhile admirers see as a desperate act, insisting that a full-scale conversion to nuclear power is the

---

[112] Frontline (PBS), "Three Mile Island: The Judge's Ruling," in *Nuclear Reaction: Why Do Americans Fear Nuclear Power?* (Apr 22, 1997).

only thing that can save the planet from catastrophic climate change. He is but the most unlikely of a considerable group of people calling for a speedy embrace of nuclear power, despite the drawbacks of nuclear waste and the remote possibility of catastrophic accidents that have made the very word "nuclear" politically radioactive for the past few decades. President Bush hasn't even uttered the word once in his two terms in office. (He's come close, though.)

The melding of environmentalism with a pro-nuclear stance now embraced by Dr. Lovelock, Patrick Moore and many others was elucidated in 1996 by Dr. Bruno Comby in his book Environmentalists for Nuclear Energy. When Dr. Lovelock wrote the foreword for the English edition in 2001, his advocacy of nuclear power created shockwaves throughout the environmentalist community. Both Lovelock and Moore, as well as many policymakers, have been influenced by Comby's work. The international organization he founded, Environmentalists for Nuclear Energy (EFN), seeks to dispel unfounded fears and the misguided notion that environmental awareness and nuclear power are incompatible. Their message is, in fact, just the opposite: that nuclear power is the solution to generating the power that renewables alone simply cannot provide.

People respond strongly to fear, even when the decisions they make under its influence would seem to be against their own best interests. If there's one thing that many people viscerally connect with the very idea of nuclear power it's fear. While it's an unfounded misconception that a nuclear plant accident could result in a nuclear explosion *à la* Hiroshima, there are of course very real concerns and serious dangers that must be confronted when considering the use of nuclear power plants.

In terms of serious damage, Three Mile Island can't compare to the 1986 accident at the Chernobyl reactor in the Ukraine. To this day the amount of radioactive material that was released into the atmosphere at Chernobyl is a matter of debate, as is its ultimate toll in terms of future cancers and other radiation-induced health effects. Thirty-one people died quite quickly from the immediate effects of the disaster, and about twenty-five more later, but a third of a million were displaced from their homes and radiation spread far and wide. The dispersal was influenced by meteorological conditions and resulted in measurable increases to the natural background radiation as far away as Scandinavia and north of the Adriatic Sea. To this day there are restrictions on certain foodstuffs throughout most of Europe, as there will likely be for years to come. Nor is the disaster site yet secure. A project to construct the world's largest movable building to cover the entire site is being undertaken to prevent further releases of radioactive material.

In terms of raw material for antinuclear hysteria, Chernobyl is a gold mine. Greenpeace came out with their own study alleging that 200,000 deaths will result from the radiation released during the accident, and uncountable other serious health effects.[113] Yet this estimate is wildly higher than two UN studies that came out in 2000 and 2005 employing many of the world's leading radiation experts.

> The 2005 Chernobyl Forum study involved over 100 scientists from eight specialist UN agencies and the governments of Ukraine, Belarus and Russia. Its conclusions are in line with earlier expert studies, notably the UNSCEAR[114] 2000 Report which said that "apart from this [thyroid cancer] increase, there is no evidence of a major public health impact attributable to radiation

---

[113] Greenpeace, "Chernobyl Death Toll Grossly Underestimated," (Apr 18, 2006).

[114] UNSCEAR, the United Nations Scientific Commission on the Effects of Atomic Radiation, is the UN body with a mandate from the General Assembly to assess and report levels and health effects of exposure to ionizing radiation.

exposure 14 years after the accident. There is no scientific evidence of increases in overall cancer incidence or mortality or in non-malignant disorders that could be related to radiation exposure." As yet there is little evidence of any increase in leukemia, even among clean-up workers where it might be most expected. However, these workers remain at increased risk of cancer in the long term.

Some exaggerated figures have been published regarding the death toll attributable to the Chernobyl disaster. A publication by the UN Office for the Coordination of Humanitarian Affairs (OCHA) entitled *Chernobyl — a continuing catastrophe* lent support to these. However, the Chairman of UNSCEAR made it clear that "this report is full of unsubstantiated statements that have no support in scientific assessments," and the 2005 report also repudiates them.[115]

The Chernobyl Forum study involved the International Atomic Energy Agency (IAEA), the World Health Organization (WHO), the United Nations Development Program (UNDP), other UN bodies and, as noted above, the governments of the areas most severely affected. While the observed incidence of thyroid cancers was perhaps due to the release of radioactive iodine from the accident, that substance only has a half-life of eight days so its effects were limited to a relatively small population, and the vast majority of those cancers were successfully treated. The study estimates that between 4,000 and 9,000 fatalities can ultimately be expected as a result of the Chernobyl accident. It should be noted, however, that these numbers are based on the Linear No-Threshold model (LNT) of damage caused by ionizing ra-

---

[115] "Chernobyl Accident: Nuclear Issues Briefing Paper 22," (Uranium Information Center, Feb 2008).

diation, for which empirical evidence is conspicuously lacking (though by its very nature it would be nearly impossible to prove) and which is a source of considerable contention in the scientific community. In point of fact, an alternative model that does have evidentiary support,[116] called radiation homesis, asserts that low-dose radiation may actually be beneficial. The use of the LNT model in this study is a clear indication of its conservatism in estimating fatalities which may, in fact, never occur.

Of course the anties accused all those government bodies and scientists of fudging the numbers in precisely the opposite direction, with some casually tossing out a 300,000 estimate, which was then — naturally — picked up in media reports. What never gets explained in these conspiracy theories, though, is what precisely all those conspirators have to gain from participating in such a massive and unbelievably effective cover-up. Meanwhile, hundreds of thousands are silently dying from coal pollution every year.

While design flaws were surely involved at TMI and especially at Chernobyl, operator error also played a role. Needless to say, while improvements in reactor design have been substantial since then, human operators can certainly be expected to be flawed in the future. The focus has thus been on passive safety systems in nuclear plants that can prevent catastrophic accidents by virtue of the reactor design itself and the physical properties of its materials. There are, however, hundreds of nuclear plants in operation around the world without the most modern passive safety systems in place, and these are aging. Realistically, old nuclear plants are more dangerous than new ones.

And yet accidents like these are only one of the strikes against nuclear power. The other major issue is with the nuclear waste that these plants produce. While those in the nuclear in-

---

[116] S. M. Javad Mortazavi (EFN), "High Background Radiation Areas of Ramsar, Iran," (Kyoto, Japan: Kyoto University, 2002).

dustry downplay the amount of waste that's been accumulating since the dawn of the nuclear age, when the numbers get substantially over 100,000 tons the general public can be forgiven for considering that as quite a bit. The fear factor comes into play even more when faced with the unthinkably long time that this material will continue to be radioactive. For all intents and purposes from a human point of view, we're talking about forever. Nobody wants to leave a legacy of nuclear waste to future generations, not even those who trust in the efficacy of the burial methods now being developed around the world.

Nuclear plants do, however, use precious little fuel to generate prodigious amounts of clean (aside from the waste) power. But many predict that uranium prices will rise in the not-too-distant future especially if many countries begin to ramp up their nuclear plant construction. That very construction, and the operation of new plants, presents a different problem. There have been very few students majoring in nuclear engineering since the decline of the industry, at least in the United States. There is a genuine concern that if many new plants are planned there won't be enough competent trained personnel to either build or operate them.

It has also been argued that electricity generated with nuclear power is much more expensive than those in the industry make it out to be. This too is a bone of contention among anti- and pro-nuclear groups. It is certainly true that nuclear plants in the United States have been expensive, not least because the resistance of anties has forced long delays and even cancellations in construction. Partly, too, it's because many different designs have been used as the technologies evolved, resulting in one-off versions of what already would have been very expensive projects. Yet even when taking these factors into account, nuclear power plants produce electricity at a very competitive rate, in part due to the very low cost of the fuel relative to the amount of power it can generate. There is often more than a bit of disingenuousness at play when calculating these figures, depending on the slant of those doing the calculations.

But if the danger of meltdowns and an eternal legacy of nuclear waste isn't enough to turn people away from nuclear power, there is the ominous potential of nuclear proliferation. This very issue is what led to the Atoms For Peace program arranging to make the USA home to much of the world's nuclear waste, since we didn't want non-"nuclear club" members having their hands on materials they could possibly use for making bombs. It is also why Jimmy Carter ordered a ban on the reprocessing of nuclear fuel, which can separate out weapons-grade material. At the time it was hoped that our example would encourage other countries with nuclear reactors to forswear reprocessing and thus lessen the risks of nuclear proliferation. Unfortunately, few considered following our lead, and in any event there were other more effective avenues still open to acquire weapons-grade material for those who were committed to doing so. The problem of nuclear proliferation refuses to go away.

If proliferation has been one of the issues people object to when it comes to nuclear power, today's light water reactors (LWRs) have been less of a concern than breeder reactors. Almost since the day the concept of peaceful use of the atom was envisioned, physicists realized that breeder reactors had the potential to provide virtually unlimited amounts of nuclear fuel. A breeder, as its name clearly implies, is able to create more fuel than it burns by exposing uranium and/or thorium to the reactor's fission process and thus creating more fissile material than the reactor is consuming. While all nuclear reactors act as breeders to some extent as the fuel undergoes a series of transformations during the reaction, breeder reactors are specifically designed to maximize this process, and have thus been shunned as possible plutonium factories. Whereas many anties decry any sort of nuclear power plants, an even larger slice of the populace (and their politicians) reflexively dismiss the use of breeders based on their proliferation potential.

Whatever one's opinion of nuclear power, though, the fact is that nuclear plants are capable of producing vast amounts of electricity without adding greenhouse gases to the environment. In the view of ever-increasing numbers of people concerned about global warming, they represent the only possibility for revolutionizing base load energy production within a time scale that can be expected to ameliorate the effects of global warming before it's too late. James Lovelock is not the only one who sees that writing on the wall.

Any other suggestions?

Our energy problems and the environmental quagmire we've created for ourselves are certainly not bedeviling us because of any lack of good intentions, the Kyoto Accords and their successors being probably their most famous manifestation. But the Kyoto Accords are not, despite the claims of their enthusiasts, the solution to anything. They've got more holes than the ozone layer. The time for half-measures has passed. We need nothing less than pollution-free primary power generation and pollution-free energy carriers, without having to resort to fantastical speculations of technological breakthroughs to bridge the yawning gaps between us and our futuristic visions. Wishing and hoping and dreaming won't make it so. And waiting for a hundred years won't either.

> Among climate scientists, a consensus has developed that we must cut projected global emissions at least in half by the year 2050. But a few leading scientists have begun to suggest that reducing pollution simply can't be done fast enough to prevent a planetwide meltdown. "This

is not a goal that can be achieved with current energy technology," says Marty Hoffert, a physicist at New York University. "I think we need to admit that and start thinking bigger."[117]

Well, Marty, I respectfully disagree about the unachievability, but I fully agree that we have to start thinking bigger. If the political will can be summoned, by 2050 we can effectively halt our regrettable contributions of global warming gases, not just cut emissions in half. But to do that, we'll have to break out of the confines of categorical thinking and consider some new directions.

---

[117] Jeff Goodell, "Can Dr. Evil Save the World?," *Rolling Stone* Nov 3, 2006.

CHAPTER THREE

A Necessary Interlude

"The greatest shortcoming of the human race
is our inability to understand the exponential function."
— Dr. Albert A. Bartlett, Physicist

S PEAKING OF A global revolution in describing what must lie in store for this generation is hardly meant to be glib. Resolution of the serious problems discussed in Chapter One — global warming, nuclear proliferation, air pollution, nuclear waste, and resource wars — will require, and will create, profound changes in societies around the world. Unlike most revolutions, however, we can engage in this one with our eyes wide open. Since it will be fairly easy to predict many of the social, political, and economic stress points, enlightened public policies can ameliorate the most problematic issues to effect a smooth transition into a greatly improved future for all of us.

Ah, but that word "enlightened" is freighted, is it not? Seen from the perspective of the United States in the first decade of the 21st century, it takes quite a leap of faith (or naiveté) to assume that policymakers are capable of even understanding what this revolution will portend. Sorely tempted to cite a few illustrative examples among the wealth of ludicrously ignorant words that have been uttered by the denizens of Washington,

I will with difficulty restrain myself to focus on the nature of the problem at hand.

The gestation period of modern science was long and painful. Just look at the story of Galileo. Yet by the time the Industrial Revolution worked its wonders (not without its own serious social and political dislocations), the general populace began to see scientific advances as something to be excited about. For most of the 20th century, the positives outweighed the negatives in the public perception of science, and new problems arising from one invention or another were often casually ignored under the blithe assumption that "the scientists will figure it out."

The advent of the nuclear age was a classic example of this mindset. Emerging from World War II, Americans[118] were flush with their political and technological success. The sky (and beyond) was the limit. When the potential of nuclear fission for peaceful purposes was glimpsed, they jumped into it with both feet. The problem of nuclear waste was definitely recognized early on by the scientists who were developing the technologies. At the time, though, environmental awareness was still in its infancy. Generally, if you had something to throw away, you'd throw it away pretty much anywhere. We're still cleaning up the messes decades later.

Those who today deride the idea that science can be the great solver of all problems can clearly be forgiven for their cynicism. Even as we benefit from amazing scientific discoveries, we unwittingly end up facing unforeseen consequences with

---

[118] The reader will please forgive me for referring generically to the inhabitants of the United States as Americans. I fully realize that North and South America include many nations and peoples, and that inhabitants of the USA have co-opted the term as if they were the only ones living there. Unfortunately it's quite difficult to use the term "United Statesians" without distracting the reader from the topic at hand. I'm a victim of an unwieldy country name. Apologies to all my fellow North, South, and Central American brothers and sisters, including my Canadian wife.

sometimes deadly implications. The development of modern medicine is perhaps the most salient example.

Until Louis Pasteur and his kind started figuring out germ theory, things were pretty much perking along in the "nasty, brutish, and short" mode. Not until the year 1800 did the population of the earth reach the one billion mark. But once modern medicine kicked in all hell broke loose. Over the course of the twentieth century the population of the world quadrupled. What this has meant for our environment, and for all our fellow non-human creatures, is profound to say the least.

Our earth is a finite sphere, and thus it is undeniable that population must remain within some sort of limits. In a world groaning under the burden of billions of people, it is simply delusional to deny the threat that overpopulation poses to our planet. Yet even with the world's population projected to increase 50% by mid-century, many of the world's most influential leaders seem oblivious to the situation.

This is illustrative of a general disconnect between scientific progress and the evolution of social consciousness. The advances of science seem to have outpaced humanity's ability to adapt. Rather than encouraging people to examine pressing issues with logic and reason, an antagonistic anti-intellectualism has taken hold of many, certainly in America at least. So we find ourselves on the horns of a dilemma. On the one hand we have the seemingly unstoppable march of science, and on the other an anachronistic mindset more suited to life in the Dark Ages. The ensuing problems are exacerbated by the sheer volume of people on the planet, and that number is rising with appalling speed, lending an urgency to our environmental problems that might otherwise be somewhat postponed. Yet who is prepared to forgo the benefits of modern medicine in order to bring the critical population portion of our dilemma under control?

This is not to say that many people wouldn't be perfectly happy — or at least willfully oblivious — to withholding modern medicine from others in geographically and culturally distant lands. Such an execrably inhumane attitude confronted me when I founded a nonprofit organization some years ago with the intention of drilling water wells in poor villages to prevent the dreadful rates of mortality from waterborne disease. I was frankly aghast at the number of seemingly normal people who, in one way or another, cast doubt upon the advisability of preventing the needless deaths of children in underdeveloped countries lest they survive to reproductive age and only add to our population dilemma.

Let it be said that those who are unwilling to forgo the benefits of modern medicine, electricity, air travel, safe food and water, and all the other fruits of technology have no right to expect others to deny themselves those same things simply by dint of their nation of birth. Indeed, the well-documented link between an improvement in standard of living and population self-control would more logically lead us to attempt to spread both education and modernity to all corners of the earth. Such a course of action would most effectively address the population growth that is arguably one of the greatest developing crises in the history of our planet. It is the height of selfishness to countenance consigning billions of people to an inferior life so that the "civilized" nations can greedily pillage the world's resources. Such a position, besides being ethically unconscionable, is based on outmoded thinking, as will be made clear in the pages to follow.

Until the peoples of the earth are willing to abandon the religious and/or cultural strictures that prevent them from limiting the size of their families, we will be faced with ever-greater demands for all the world's resources. Demographers predict that well before the end of this century humanity will have expanded its ranks — barring major disasters of some kind — to over ten billion people. And all of those people will need energy.

It would be deluded at best to pretend that energy conservation and self-denial are going to make a dent in this problem, yet that is about as deeply as some environists[119] are thinking. And it is every bit as foolish to believe that we can dramatically increase the world's population while we maintain the fossil fuel power model. Self-denial is not a policy. Neither is denial.

Both directly and indirectly, science has combined with human folly and shortsightedness to create a critical mess. And like it or not, we'll have to turn to science to deal with it. Neoluddites will have to be kicked to the curb, though hopefully most of them will be able to open their eyes to reality and become part of the solution rather than remaining part of the problem. Likewise the fossilized thinking of the fossil fuel forever advocates must be abandoned, and not a moment too soon. Indeed, we can only hope it's not too late.

As I wrote at the outset, revolutions tend to upset people, as Louis XVI, Marie Antoinette, and countless others throughout history have discovered to their chagrin. But nobody need lose their head over this one. Yet this new revolution, while peaceful, will just as surely upset people on both the left and the right ends of the political spectrum. That's both regrettable and inevitable. Regrettable because it's difficult to convince politicians to make tough decisions when they know that a portion of the electorate will have to be dragged kicking and screaming into a new paradigm. The status quo, and perhaps especially the energy portion of the status quo, is nearly immovable. It will take a tremendous amount of pressure from the populace before our policymakers are ready to embrace solutions, even if they themselves believe that the solutions are ultimately beneficial. Most politicians like their jobs, have kissed a lot of babies (etc.) to get where they are, and are more than willing to temper their good sense with political expediency when the former might lead them to electoral ruin. Yet time is not on our side in this instance.

---

[119] Environmentalists in whom the "mental" portion is substantially inoperative.

My hope is that you, dear reader, can set aside for a moment your own fears, preconceptions, categorical thinking, and scientific and political biases to consider my proposal for a solution to the seemingly intractable problems that I promised to deal with in the beginning of this book. You will not be asked to make any great leaps of faith or technological fantasy. Amazing as it may seem, the technologies to solve some of the greatest challenges of our time are well in hand. First, though, you — and many like you — have to be convinced. Only then can our decision-makers possibly be persuaded to set aside their deadly inertia and take the bold steps necessary to implement real solutions.

CHAPTER FOUR

Newclear Power

"Mankind does have the resources and the technology to cut greenhouse gas emissions. What we lack is the political will."
— Dr. Paul Davies, Physicist

O F ALL THE energy systems we've discussed (albeit briefly, of necessity), the one with the greatest potential for reducing the threat of global warming is arguably nuclear power. It's been surprising to see the range of individuals who have embraced this option, who have pronounced themselves willing to settle for more and more nuclear waste and the widely feared dangers of proliferation and possible (though unlikely) accidents. It is a measure of how deeply concerned they are about global warming. But there is still a vocal opposition to nuclear power by those who aren't ready to discount the negatives that seem to be an inextricable part of the package.

Opponents of nuclear power might assume that those who support it are being dismissive of such concerns either out of foolish heedlessness or desperation. But from the beginning of the nuclear power era the physicists, engineers and others who worked at the cutting edge of that research recognized both its promise and its shortcomings. Having identified the most serious problems, they set out to solve all of them, determined to leave no loose ends.

Argonne National Laboratory (whose western branch was recently rechristened Idaho National Laboratory) was the focus for America's nuclear power research and development since the beginning of the nuclear age. In 1964 a research reactor called the EBR-II was built to demonstrate a breeder reactor system with on-site fuel reprocessing and a closed fuel cycle. During the thirty years of its operation, many advances were incorporated into its design and proved eminently workable. The project was a resounding success. The advantages of such a system are so far superior to the light water reactors (LWRs) now in use that one might be forgiven for wondering why this technology has not completely supplanted current systems.

There is much here to wonder about. Why was the program suddenly terminated in 1994, just one step shy of its full demonstration of proof of concept, after thirty years of flawless performance? Cost was not an issue, especially since the Japanese had offered to chip in $60 million to finish the research. It seems especially ironic to see Al Gore today as the leading light in the climate change field when it was the Clinton administration (with Gore as vice president) that urged Congress to shut down the EBR-II. There were certainly no technical or economic reasons to do so.

There has been speculation that Clinton was bowing to antinuclear political pressure and that the shutdown was a payback for the support of environmentalists. Certainly Clinton and Gore's 1992 campaign stressed renewable energy development and a distinct lack of support for nuclear power. It has also been suggested that Clinton's choice as Secretary of Energy, Hazel O'Leary, would have been wary of the threat to the fossil fuel industry that the Argonne project represented, having previously been a lobbyist for fossil fuel companies. She and Senator John Kerry led an impassioned opposition to the project, arguing that it represented a proliferation threat. Since the

EBR-II's design was specifically intended to reduce proliferation risks, however, their opposition would seem to be a case of either ignorance or duplicity. It seems entirely believable that the shutdown of the program was ultimately due to misinformation and misunderstanding of the legislators who voted to kill it. It's been said that they didn't understand the difference between PUREX fuel reprocessing (which does present a proliferation threat because it isolates weapons-grade material, albeit of poor quality) and the proliferation-resistant fuel recycling that was intended to be an integral part of the new reactor system. The Senate, in fact, didn't go along with Clinton's recommended program termination, but the House prevailed in conference committee and the program was killed. (You can read the whole deplorable story in Chapter 12.)

If the Integral Fast Reactor (IFR) concept that the EBR-II represented was so far superior to current designs, you may wonder why the information hasn't made its way out into the public arena since 1994. When I first began to research this technology in 2001 I found even the people who worked at Argonne quite reticent to discuss it openly. After a considerable amount of communication I finally asked one day why the person who was my source of information there wasn't more forthcoming. It seemed I always had to pry information out of him a piece at a time. Finally he told me that the Department of Energy had issued a directive that the technology was not to be publicized. I could have specific questions answered, but I would have to figure out what those questions would be.

What was doubly ironic is that the chief engineer for the EBR-II project, Leonard J. Koch, was awarded the prestigious Russian Global Energy International Prize by Vladimir Putin in June 2004 for his work on the project.[120] And this was

---

[120] Argonne National Laboratory News Release, "Argonne Fast-Reactor Pioneer Receives International Prize," (May 7, 2004).

happening at a time when the Argonne people were under a virtual gag order to prevent free discussion of the project in their own country! What could have prompted the U.S. Department of Energy, then under the watchful eye of Spencer "I-never-met-a-gas-hog-I-didn't-like" Abraham, to squelch publicity about this promising technology? Why, indeed, have both Democratic and Republican administrations thrown bars in the wheels of their own scientists who'd worked for over thirty years — with stunning success — to develop and demonstrate an incredibly promising energy technology?

Rather than venture into areas ripe for speculation, I will leave my readers to draw their own conclusions and, hopefully, ask themselves and their political leaders some penetrating questions. In order to encourage that, a description of the technology, and what it could mean to our planet in its current dire straits, will be presented here. I will make every attempt to refrain from overly technical descriptions. Footnotes will be provided for those who wish to delve further into the details, and the glossary at the back of the book provides descriptions of the terminology and acronyms.

If you find this to be tough sledding despite my efforts to the contrary, please don't be dismayed. A cursory understanding of the basic concepts is helpful, but you need not be concerned if the details escape you. The salient points will be made quite clear regardless, as the book progresses. I would, however, mention one exception to this. If you happen to be a person for whom anything with the word "nuclear" in it is anathema and you still feel that way even after you get to the end of this book, then it would behoove you to make sure that you do, indeed, understand these principles. If not, I would submit to you that you don't have a sufficient basis upon which to espouse a dogmatic position. If you cling to the belief that nuclear anything is necessarily bad (excepting, perhaps, the nuclear family), and yet don't understand how it works, then you're probably just accepting it on faith from someone who may be as ignorant about the facts as you are. Even worse, you may have placed your trust in someone who knows better but who preaches an anti-nuclear ideology for reasons that are either self-serving or willfully blind to the facts, in which case deliberate distortions and outright lies are unfortunately not at all uncommon. In any event, at this point I would implore you to withhold judgment until the evidence is presented. I suspect you may be both surprised and hopeful when all is said and done.

Nuclear Physics 101

The process that powers nuclear reactors today is termed fission, and uranium is the basic fuel. In a reactor, neutrons[121] are naturally released from fissioning atoms and collide with the nuclei of other atoms in their vicinity. The absorption of a neutron often causes the nucleus of the impacted atom to split apart (fission), thus creating *fission products* — isotopes[122] of two new elements of about half the mass. In the process of splitting, the impacted atoms themselves release neutrons, which continue the process by colliding with more atoms, causing a chain reaction that is harnessed for the heat it

---

[121] A neutron is a subatomic particle with no net charge. Both protons (with a positive charge) and neutrons are found in the nucleus of all atoms (except for hydrogen's protium form). Electrons, carrying a negative charge, reside outside the nucleus.

[122] The number of protons in an atom's nucleus determines which element it is, but a variation in the number of neutrons in an element's nucleus is what the term isotope denotes. Thus an element is distinguished by its name or its chemical symbol, while a number following it designates its "mass number," the total number of protons plus neutrons. Thus U-238 denotes the most common isotope of uranium, which has 92 protons, with a total of 146 neutrons (238 − 92). U-235, with 3 fewer neutrons, is needed to fuel nuclear reactors and can also be used to build nuclear weapons.

produces. The reaction is controlled by materials that either slow down or safely absorb neutrons, keeping the heat within tolerable limits and thus preventing the fuel from melting. Fluid coolant is piped around the fuel to draw off the heat and harness it via a heat exchanger to run a steam turbine that powers an electricity generator.

This can be thought of as modern alchemy, where one element is transformed into others. But whereas the alchemists of old seemed intent on producing one particular element — gold — the fission process is considerably more random, producing a great variety of elements. (And fission works, while ancient alchemy didn't.) Some of the resulting isotopes are stable, but almost all of them are decidedly unstable at first, and spontaneously emit (or radiate) subatomic particles as they decay towards a stable condition.

In some of the fuel atoms, absorption of a neutron does not lead to fission, but to the creation of a slightly heavier isotope. Some of these newly created heavy elements are themselves good producers of neutrons when they split, and thus contribute to furthering the chain reaction. Many of the lighter elements that result from the splitting of atoms, however, impede the reaction by absorbing neutrons, thus acting as so-called nuclear poisons.

It is the buildup of the nuclear poisons that is the limiting factor in the usability of nuclear fuel. The types of nuclear reactors in commercial use today operate with relatively slow neutron speeds, which increases the cross-section for absorption of neutrons and thus the probability that fission will occur. The usual fuel of choice is uranium, with the concentration of the minor isotope, U-235, enhanced (the fuel is "enriched"). Virtually all current reactors use water to slow ("moderate") the neutrons (lowering their kinetic energy) and to carry off the heat. Hence such reactors are generically classified as "thermal" reactors. Nuclear poisons build up eventually and make further reactivity impossible. The fuel is then removed from the reactor and either discarded as nuclear waste or, in some cases (though not in the USA), destined for partial recycling.

The Integral Fast Reactor (henceforth IFR), as might be deduced from the word "fast" in its name, is a type of reactor that allows the neutrons to move at higher speeds by eliminating the moderating materials used in thermal reactors. The greater velocity of the neutrons results in more energetic splitting and thus a greater number of neutrons being liberated from the collisions. The result is that the fuel is utilized much more efficiently. Whereas a normal nuclear reactor utilizes less than one percent of the fissionable material that was in the original ore, with the rest being treated as waste, a fast reactor can burn up virtually all of the uranium in the ore.

That quantum leap in efficiency is only the tip of the iceberg, though. For the fuel can then be recycled on-site in a process that removes the fission byproducts and incorporates the actinides[123] from the used fuel into new fuel rods, which are then reloaded into the reactor. The fission products, being the ashes of the process, if you will, are not usable as fuel (or as weapons). They can be stabilized by vitrification, a process that transforms them into a glasslike and quite inert substance for disposal. In this form they can be stored for thousands of years without fear of significant air or groundwater contamination.[124]

Yet the waste coming from an IFR doesn't have to be stabilized for anwhere near that long. Unlike the "waste" from the thermal reactors in use today, the waste elements from an IFR

---

[123] The 14 chemical elements that lie between actinium and nobelium (inclusively) on the periodic table, with atomic numbers 89-102. Only actinium, thorium, and uranium occur naturally in the earth's crust in anything more than trace quantities. Plutonium and others are heavier, man-made actinides resulting from absorption of neutrons.

[124] D. H. Bacon & B. P. McGrail, "Waste Form Release Calculations for the 2005 Integrated Disposal Facility Performance Assessment," (Pacific Northwest National Laboratory, July 2005).

have much shorter half-lives[125] than the actinides that have been retained in the reprocessing and subsequently reloaded into the IFR's core for further fissioning. The nuclear waste problem, probably the most common concern about nuclear power, is seen as serious primarily because of its long-lasting radioactivity, for some of the actinides remain appreciably radioactive for thousands of years. With the actinides removed from the spent fuel, dealing with this new type of nuclear waste becomes quite manageable. Though very radioactive (a shorter half-life means more intense radioactivity), the vitrified IFR waste can be placed in lead-lined stainless steel casks and safely stored on-site or transported for storage elsewhere. Within a few hundred years — millennia before there is any degradation of the vitreous mixture that locks it in — the radioactivity will have diminished to below the level of naturally occurring ore. And unlike the actinide-containing waste, no weapons-usable materials are involved.

Yet there is an even better feature of IFR fuel than its relatively benign waste. For the new actinides used to augment the spent IFR fuel during its reprocessing can come from the nuclear "waste" from thermal reactors, which we are all so concerned about. Plutonium and uranium from decommissioned nuclear weapons can also be incorporated into fast-reactor fuel. Thus we have a prodigious supply of free fuel that is actually even better than free, for it is material that we are quite desperate to get rid of. Uranium, plutonium, and other actinides, both weapons-grade and otherwise, will go into the IFR plants. Only non-actinides with short half-lives will ever come out. We will eliminate the problems of both radioactive longevity and the potential for nuclear proliferation.

---

[125] Half-life refers to the amount of time required for a radioactive substance to decay to half its original quantity. As radioactive elements decay toward a stable state their radioactivity decreases, eventually passing below the harmless levels of normal background radiation.

Which brings us to one likely reason why fast reactor technology has been ignored all these years. Because fast reactors are capable of creating more fissile material than they burn, they are known as breeder reactors. And because breeder reactors create plutonium, they have been a special target of anties and politicians concerned about proliferation. As in so many issues having to do with nuclear power technology, most of the resistance is due to ignorance of the technology and a generalized fear of all things nuclear.

Let's be clear about one thing: all uranium-fueled nuclear reactors create plutonium. Here's how it works: When uranium ore is extracted from the ground and milled, it contains about seven-tenths of one percent uranium 235 (U-235). This is a fissile material, meaning that it is so prone to splitting when it absorbs a neutron that it will maintain a fission chain reaction if enough of it is brought together in the same place. The other 99.3% of the uranium is made up almost entirely of U-238, which is not fissile but fertile. Fertile materials are those that do not readily fission in a neutron flux, but which, upon absorbing a neutron, are transmuted into fissile isotopes. A handy rule of thumb when discussing actinides is that if the mass number is even, they're fertile. Odd numbered actinides, on the other hand, are fissile.

In most thermal reactors, uranium must have a higher concentration of fissile material than its natural concentration of 0.7%, so it is put through an enrichment process to boost its percentage of U-235 to about 4%. Once this concentration is attained the fuel can be assembled into a critical mass, the amount necessary to maintain a fission reaction. The U-238 that makes up the other 96% of the fuel is then bombarded with neutrons as the fission proceeds since it is, of course, in the neighborhood.

When a neutron hits an atom of U-238, one of two things can happen. Either the atom fissions (unlikely) or it absorbs the neutron in a process known as neutron capture. You'll remember

that neutrons have no charge, whereas protons (their compan-
ions in the atom's nucleus) have a positive charge and electrons
have a negative charge.

Once U-238 absorbs a neutron it would be expected to become the radioactive isotope U-239, and it does. But U-239 has
a half-life of just minutes, so it quickly undergoes beta decay,[126]
becoming a different element, neptunium 239. But Np-239 has
a half-life of only 2.35 days, so it also soon undergoes beta decay.
Now the atom of U-239 has gone from having uranium's 92 protons to 94 protons (through 2 consecutive beta decays). Since the
number of protons determines the identity of an element, it is no
longer uranium, nor neptunium. It is plutonium (Pu-239).[127]

So now you have two fissile elements in the reactor core:
what's left of the original 4% of U-235, plus some Pu-239, which
itself begins to fission. The neutrons being liberated from both
these elements continue to not only produce fission products, but
also to create more plutonium from the remaining U-238, thus
sustaining the reaction longer than would be the case without
the creation of plutonium. By the time the buildup of nuclear
poisons necessitates the removal of the nuclear "waste" there's a
considerable amount of plutonium that's been created. A nor-
mal-sized nuclear power plant of one gigawatt capacity—sufficient to power about a half million European homes, but only
about half that many in the more power-hungry USA—will
expel nearly 500 pounds of plutonium in its spent fuel over the
course of a year.

---

[126] In beta decay, a neutron is converted into a proton while emitting an
electron and an anti-neutrino. Don't worry, there won't be a quiz on this.
The point is, a proton replaces a neutron, thus changing one element into
another.

[127] Since uranium had been named after the planet Uranus, the discover-
ers of the next two elements named them in ascending order after the last
two planets of the solar system. That was back in the good old days, of
course. (Sorry about your recent demotion, Pluto!)

In both thermal and fast reactors, the plutonium produced is intimately mixed with a large amount of U-238 and other elements, and the spent fuel would have to be reprocessed in order to get pure plutonium. This can be done as easily with irradiated fuel from an ordinary thermal reactor as it can from the "breeder blanket" of a fast reactor. So the hue and cry about the proliferation dangers of breeder reactors is actually much ado about nothing special. The danger of nuclear proliferation isn't an issue of thermal reactors vs. fast reactors; it's an issue of maintaining tight control over the entire nuclear fuel cycle, regardless of the type of reactor. One of the great benefits of the IFR over thermal reactors is that the reprocessing facility is located in the same complex as the reactor itself—hence the "Integral" in "Integral Fast Reactor" (IFR). In an IFR plant, all actinides—including plutonium—are kept sequestered in an extremely radioactive environment while they are repeatedly sent through the fast reactor until they are transformed into energy.

The so-called pyroprocessing that occurs at an IFR site is quite unlike the PUREX (**P**lutonium and **U**ranium **R**ecovery by **EX**traction) reprocessing, which isolates weapons-purity plutonium from a thermal reactor's spent fuel. During the entire relatively simple pyroprocess within the confines of the IFR, the plutonium is always in combination with elements that make it impossible to use for weapons without further, PUREX-type processing, and is so radioactive that the entire operation is done remotely behind heavy shielding. Once the new material that we want to dispose of is added from outside, it too is removed from possible weapons use once and for all. Thus all the actinides in spent fuel from thermal reactors, as well as weapons-grade material we wish to get rid of, can be sent to IFRs. Instead of being a plague on future generations, the energy potential of the actinides is fully utilized in the production of electricity.

Consider, if you will, what this means in terms of energy

availability. Nuclear "waste" — which in today's terms can now be seen to be a gross misnomer — from LWRs[128] still contains about 95% of the fuel's original energy. IFR plants can burn, in time, *all* of the actinides that have been mined, not just those that make it into the LWR's fuel. None of the actinides that enter the site will ever leave it, until the time comes that all the plutonium from thermal reactors has been used up, and excess fissile material must be bred and transported to new reactors that need an initial loading. As we'll see later on in the book, for all the worry about the long-lived nuclear waste building up all over the world, we can easily use it all up in IFRs. And once it's all used up, all we'll need to keep the then-existing IFRs operating is U-238, the principal component of depleted uranium (DU), which is a byproduct of uranium enrichment and the main component of all reactor fuels.[129] We have so much of this already available that it could provide all the power needs of the entire planet for hundreds of years before we need to mine any more uranium. This is the same depleted uranium that is currently being used in both defensive and offensive weaponry, primarily by the United States. It would be a great improvement if we'd use it for generating electricity instead of shooting it at people.

Let us not forget the hazards and environmental insult of uranium mining and milling, which is a constant and ever-growing requirement of thermal reactors. Once all the thermal

---

[128] LWR: Light-Water Reactor. A thermal reactor that is moderated and cooled by ordinary water, which must be fueled with uranium that has been enriched to about 4% U-235. Most of the world's power reactors are LWRs, but by no means all. Some, moderated by heavy water or graphite, can use natural uranium. In fueling LWRs, some 85% of the ore's energy is left behind in the tailings from the enrichment process, and only about 5% of what makes it into the fuel gets consumed.

[129] All current reactors, that is. Thorium is another possible reactor fuel, but as yet the technology is not mature.

reactors reach the end of their useful lifetimes and are all replaced with IFRs, the world's uranium mining and milling operations can be completely shut down for centuries. Likewise all uranium enrichment facilities will be obsolete, as will large, centralized plants for reprocessing spent fuel from thermal reactors.

Once that point is reached, all it would take to keep a one-gigawatt reactor running would be about a milk-crate quantity of depleted uranium every three months. And if the stuff wasn't so ungodly heavy (1.6 times as dense as lead), it's safe enough that a person could just carry it into the plant by hand. Except for weapons-grade plutonium possessed by nations in the "nuclear club," none would ever be in existence outside the IFR plants.

But whereas nuclear waste and proliferation are serious problems that can be rectified with IFR technology, what about the possibility of nuclear accidents? Once again the IFR design has proven to be a stellar solution. One of the major problems with thermal reactors is the fact that they use pressurized systems for their coolants. Both the Three Mile Island and the much more serious Chernobyl accidents were due to coolant problems, faulty readings from monitoring devices, and operator error. In addition, the antiquated Chernobyl didn't even have a containment building, thus allowing the release of radioactive substances that was prevented in the case of Three Mile Island.

The physicists and engineers who designed the IFR wanted to eliminate even the remote possibility of accidents by using passive safety, which relies on the inherent physical properties of the reactor's components to shut it down in even the most adverse situations. And once again they figured out how to do it.

The reactors in an IFR complex are often referred to as LMRs, meaning Liquid Metal Reactors (or sometimes ALMRs, Advanced Liquid Metal Reactors). The Argonne project used a large vat of liquid sodium in which the reactor vessel itself was immersed. Sodium has the advantage of being an excellent conveyor of heat, as well as innate characteristics that prevent it from interfering in the fission process. A closed loop of sodium passes through this pool, transporting heat from it into a separate area where it boils water in a second heat exchanger. The (non-radioactive) water, as with thermal reactors, is thus converted to steam for generating electricity with the plant's turbines. The now-cooler sodium in the heat-transfer loop circulates back through the reactor pool heat exchanger in a continuous process.

Unlike thermal reactors, however, this is all done at atmospheric pressure, or nearly so. The closed loop utilizes a low-pressure pump just sufficient to maintain the sodium flow, moving cooled sodium from the heat exchanger back into the reactor area. There is also a small circulating pump immersed in the main tank to transfer heat more efficiently from the reactor core to the sodium pool. The diagram[130] shows an IFR that incorporates a breeder blanket of fertile material (U-238) that is being converted to fissile material, to "breed" more nuclear fuel. In the beginning of the conversion to IFRs, new reactors would be fueled with actinides from used thermal-reactor fuel. For the most rapid growth of nuclear power, IFRs would be loaded to breed the maximum possible amount of new fissile material, using the excess to start up new IFRs. Should the time come when no more generating capacity is needed, the reactors could be operated in the "break-even mode," to simply maintain the plant's own operation with the breeding reduced to a subsistence level.

The IFR concept considerably simplifies the entire nuclear power system, utilizing far fewer valves and pumps than even the most advanced thermal reactors and avoiding the potential problems of high-pressure coolants. The metal fuel, unlike

---

[130] Illustration courtesy of Andrew Arthur

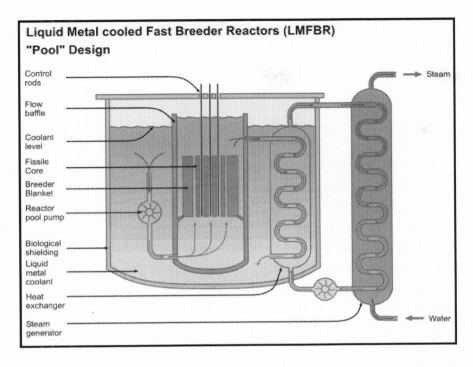

the ceramic pellet fuel of thermal reactors, conducts heat much more efficiently and is thus able to dissipate it far more effectively. The fuel pins' unique composition is such that if they begin to overheat the resulting expansion decreases their density to the point where the fission reaction simply shuts itself down.

The passive safety characteristics of the IFR were tested in EBR-II on April 3, 1986, against two of the most severe accident events postulated for nuclear power plants. The first test (the Loss of Flow Test) simulated a complete station blackout, so that power was lost to all cooling systems. The second test (the Loss of Heat Sink Test) simulated the loss of ability to remove heat from the plant by shutting off power to the secondary cooling system. In both of these tests, the normal safety systems were not allowed to function and the operators did not interfere. The tests were run with the reactor initially at full power.

In both tests, the passive safety features simply shut down the reactor with no damage. The fuel and coolant remained within safe temperature limits as the reactor quickly shut itself down in both cases. Relying only on passive characteristics, EBR-II smoothly returned to a safe condition without activation of any control rods and without action by the reactor operators. The same features responsible for this remarkable performance in EBR-II will be incorporated into the design of future IFR plants, regardless of how large they may be.[131]

These worst-case scenario trials were meant to account for the most serious possible circumstances such as devastating earthquakes or meteor strikes, though since that date the possibility of airliner strikes might also be added to the list of conceivable disasters. The potential problem to be prevented is overheating due to the sodium coolant being drained or lowered to the point where the fuel would be exposed. In order to avoid this, the containment structure can be built with no openings whatsoever below the level of the top of the sodium vessel. The primary vat is half-inch-thick stainless steel. A second stainless steel vessel surrounding that is designed to contain the sodium in the highly unlikely event that the primary one should spring a leak. Outside that second vessel is a six-inch thick sodium-resistant hardened concrete barrier, resting against the solid wall of the containment building, which forms the fourth level of assurance. Outside, the containment building can be banked with earth at least to that level, forming even a fifth level of redundancy. By supplying sufficient sodium to allow for maximum leakage all the way to the earthen barrier while

---

[131] From "The Unofficial IFR Home Page," which served to bypass the DOE gag order for years to keep the story of the IFR available on the internet. It disappeared in 2007. Sorry, DOE, the cat is out of the bag.

still keeping the reactor covered, a loss of coolant accident would be a virtual impossibility.

But sodium is known to be a somewhat dangerous substance in its own right, subject to easy combustion in air and explosive combustion if it comes into contact with water. In order to prevent contact with air the entire covered pool area is itself covered with a blanket of argon gas, which is nonreactive with sodium and forms an effective barrier. Being heavier than air, it is unable to escape from such an area since there is no egress below the top of the tank. Argon is also used in the pyroprocessing facility where the fuel is recycled, though in that process sodium is not involved. The only other possible contact with air or water for the sodium is in the unlikely event of a breach of the water side heat exchanger loop, which is constructed of double-walled stainless steel. Should a leak occur, sodium would at most flow out at a low rate because of the unpressurized system. To get an explosive reaction in air you need atomization, which isn't an issue in an unpressurized system.

Though sodium is highly reactive with air and water, it is completely nonreactive with stainless steel. When cameras were run into the double-walled sodium loops after thirty years of use in the EBR-II to check the extent of corrosion, the welders' original markings were still visible on the joints that had been welded, as they were in the tank itself when the pool was drained. In point of fact, sodium is frequently used in industrial processes because of its superb heat transfer characteristics, and one would be hard-pressed to find an incidence of a serious sodium fire. The room where the heat exchanger brings the sodium loop and the water loop together could also be filled with argon as a precautionary measure, argon being noted for its fire extinguishing (or in this case, preventive) properties. The chances of a water/sodium contact are extremely remote, considering the lack of corrosion between the sodium and stainless steel and the well-known minimal interaction between stainless steel and water. Keeping the water reasonably clean and nonreactive would be sufficient to deal with any sort of corrosion issues preemptively. During the lifetime of a plant it is unlikely that anything would have to be replaced. Based on past experience with nuclear plants (and other industrial facilities), however, the wisest course of action will be to make sure that the plant design will allow for replacement of any components that might become compromised, even if the chances of such contingencies are slim. And in a worst case scenario where the sodium and water met, it would happen in a separate building, isolated from the reactor core and its pool of sodium. No radioactive material would be involved, and the argon would smother any fire.

Though terrorism had always been a safety consideration even before its recent prominence in the public consciousness, there are several design features that can be utilized to make the reactor complexes essentially terrorist-proof. As with the EPR, the containment building can be built to withstand a direct hit from a fully fueled jetliner. A web of heavy cables can be suspended like a net above the containment and control structures, which would preemptively shred any incoming aircraft even before it made contact. But even better than that would be to simply mound earth over the critical structures once they're built, effectively keeping them above the water table but nevertheless taking advantage of the structural impregnability of massive amounts of earth.[132] Building such a structure with its sole ingress being via blast doors would make it virtually impervious to terrorism of any kind.

This tub within a tub redundant safety system provides a perfect opportunity for multiple sets of shock absorbing mounts in the event of a major earthquake. One could hardly envision

---

[132] It might be advisable to make sure that the reactors are built at least 50 meters or so above sea level, just in case the most pessimistic global warming scenarios come to pass despite our best efforts.

a scenario under which the three levels of primary containment would be breached, much less the earth itself outside them. If theoreticians and statisticians and materials scientists feel that the system as described here is still too risky (highly unlikely, but then again I'm not a statistician), how many layers of containment would be needed to make it acceptably safe, to make it one in a million safe, or one in a billion? A third stainless tub? A thicker or completely separate additional reinforced concrete barrier? Fine, no problem. Build it in. The safety factor built into the fuel rods themselves is based on the laws of physics, which are fairly immutable at this level. Eventually it gets down to the point of irrational paranoia, beyond which nothing would ever be built and we'd still be living in caves (Look out, a stalactite might fall on your head!).

Just as an incredibly improbable thought experiment let's imagine the earth suddenly yawning open and swallowing the entire complex, the sodium pouring out and catching fire in an onrushing flood that just happened to occur after the earth crushed the reactor pool to pop its top. Such yawning, gulping, then crushing scenarios are favorites of cheesy disaster movies, but unknown in real life. Of course the chance of any such event occurring anywhere, much less precisely at a reactor core, is astronomically improbable. Nevertheless, imagine tons of uranium and even some plutonium being thus inexplicably liberated in a scenario as unlikely as Elvis and Marilyn rising together from the dead. If all the earth's electrical supply were provided by IFRs and this happened by some miraculous event, the damage to humanity would still be far less serious than what our current energy systems are doing every day, with tons of polluting gases pouring from coal-fired power plants, while their soot alone kills well over half a million people per year.[133] The safety factors that would be built into the IFR plants as a

---

[133] Jeff Barnard, "Researchers Track Dust, Soot from China," *Boston Globe* Jul 13, 2007.

matter of course will most certainly provide a level of safety that will be a vast improvement over current energy systems, be they coal, oil (with its long list of disasters both large and small), gas (likewise), or even hydro power.

No matter how safe a system is, those who seek to find fault with it will often contend that a disaster is only one human error away, and that there's no way around it. That same argument will undoubtedly be leveled at the IFR system, yet it would be wildly off the mark. One of the wonders of the passive safety of IFRs is that they substitute the very laws of physics in place of human competence and mechanical performance. Rather than relying on pumps never failing (or on redundant backup pumps and systems), or on the competence of the plant operators, IFR design relies on unchanging physical laws. The boiling point of sodium is not going to change. And the temperature beyond which the fission reaction cannot sustain itself—less than the aforementioned boiling point of sodium—is likewise a function of the laws of physics. Human error cannot change these immutable conditions.

Proposing a complete replacement of fossil fuel power plants worldwide with a massive building project of IFR reactors would seem outlandish if it were to be based on the single experience at Argonne, however spectacular that program may have been (and it was). But the Americans were not the only ones experimenting with breeder reactors in the latter half of the twentieth century.

> For three decades, several countries had large and vigorous fast breeder reactor development programmes. In most cases, fast reactor development programmes were at their peaks by 1980. Fast test reactors [Rapsodie (France), KNK-II (Germany), FBTR (India), Joyo (Japan), DFR (United Kingdom), BR-10, BOR-60 (Russian Federation), EBR-II, Fermi, FFTF (United States of America)] were

operating in several countries, with commercial size prototype reactors [Phénix, Superphénix (France), SNR-300 (Germany), Monju (Japan), PFR (United Kingdom), BN-350 (Kazakhstan), BN-600 (Russian Federation)] just under construction or coming on line.[134]

A combination of factors led to the termination of these programs, not the least of which was the political pressure brought to bear by antinuclear activists. There were also a few accidents which, while not resulting in any danger to the populace, were seized upon by nuclear power foes to create political calamities. The accidents that did occur resulted from designs flaws that were eliminated in Argonne's EBR-II.

Those who conceived and built these plants understood full well that the future might present a very different political landscape and that someday this type of reactor might be necessary, whether from a diminishing supply of uranium or because of unforeseen developments. Global warming, of course, is probably the most surprising development, at best only dimly imagined in the early days of nuclear power research. Fortunately, the commitment to advancing this technology resulted in an international effort to create a shared pool of knowledge. Over forty years of fast reactor development worldwide represents a total of 300 reactor-years of operation. In the view of nuclear experts from around the world who know it from experience, this technology has reached a mature stage and is fully ready for commercial application. In fact, a fast reactor is currently under construction in India at the time of this writing, with others on the drawing board in various countries.

I once asked one of the directors at Argonne National Laboratory how the physicists and engineers felt about being ordered to essentially keep their work out of the public eye.

[134] IAEA, "Operational & Decommissioning Experience with Fast Reactors," (Cadarache, France: Mar 11-15, 2002).

He told me that from what he could tell most of them seemed surprisingly sanguine about it, assuming that global warming politics and energy supply realities would eventually trump fossil fuel politics, at which time their system would become the obvious choice. It seems that time has arrived. A first step must be a revelation of the existence of such technology to the public at large, for the implications of a worldwide conversion to IFRs are staggering. Tremendous pressure will have to be brought to bear on our political leaders in order for them to abandon the current state of affairs and strike out on a path that puts the earth, and their constituents' interests, before the interests of the giant corporations that today have a stranglehold on energy production—and, to an appalling degree, on politicians—around the world.

Before we go down that road though, there are other roads to consider: the roads we drive on. Even if all the nations of the earth agreed to rely on the far superior IFR technology for their electrical generation, we still need energy carriers of some kind for use in our automobiles and other applications where electricity is inconvenient or unavailable. Virtually every discussion I've seen of fast reactors envisions using them to produce hydrogen for the supposed future "hydrogen economy." Indeed, the recent direction of nuclear power research has been directed toward the development of high-temperature reactors specifically for the production of hydrogen for transportation. Yet we've seen that hydrogen has immense technological hurdles to surmount before it can be economical and safe. Joseph Romm, earlier mentioned as the former Clinton administration energy official and author of *The Hype About Hydrogen,* comments:

> People view hydrogen as this kind of pollution-free elixir. That all you have to do is put hydrogen in something, and it's no longer an environmental problem, which is just absurd. In fact, if you take hydrogen from fossil fuels and

run them in an inefficient internal-combustion engine vehicle, you end up with a vehicle that just generates a great deal of pollution.

People need to get out of their heads [the idea] that there is something that is inherently good for the environment about hydrogen. If you run it through a fuel cell, you have zero tailpipe emissions. We all would like zero tailpipe emissions. If you burn it, however, you don't get zero tailpipe emission, in fact. You get a lot of nitrogen oxide, because it tends to burn at a high temperature...

...The current costs of the fuel cells are about 100 times the cost of internal-combustion engines. Right now, they cost hundreds of thousands of dollars apiece. And getting them, frankly, to be within a factor of 2 of a regular car will be a stunning scientific achievement. I'm not expecting that to happen for at least two decades.[135]

If the Cal Tech researchers who predict ozone layer destruction are correct, even if all these challenges are met hydrogen may still be too hazardous to our planet to deploy as a worldwide source of fuel or, more precisely, as our primary energy carrier. But don't despair. There is a far better idea than hydrogen.

---

[135] Romm, "Just Say No, to Hydrogen."

CHAPTER FIVE

The Fifth Element

With apologies to Bruce Willis

T HE SEARCH FOR solutions to our dependence on fossil fuels has gone down some strange roads, so it probably was inevitable that it would lead to the junkyard eventually. America is a notoriously throwaway culture. It's not too much of a stretch to think that some of what we're tossing out might be worth another look.

A researcher at Oak Ridge National Laboratory in Tennessee, Dave Beach, perhaps was thinking along these lines when he came up with the idea of grinding up the metal in our nation's scrap yards and burning it for fuel. But wait a minute, metal doesn't burn. Or does it? We all know metal can get really hot and melt, but even at blast furnace temperatures it doesn't burn. Another pipe dream?

Not so, says Beach. His team has applied for funding to build a prototype car that will burn metal as fuel. It turns out that when metal is ground exceedingly fine the resulting nano-grains become highly reactive, at which point they can be ignited and will burn quite readily. The fact that they burn at a relatively low temperature results in a reduction in the emissions of carbon dioxide, nitrogen oxides and particulates, which are formed mainly at the high temperatures in internal combustions engines. The bulk of the exhaust is mainly metal oxide. So burning steel produces rust (ferrous oxide), and if that rust is heated in a hydrogen or carbon monoxide environment the oxygen will gladly abandon the steel, which can then be used again, ad infinitum.

Unlike fossil fuels, metal fuels are not really energy sources. They, like hydrogen, are energy carriers. The good thing about fossil fuels is that we can extract them from the ground and take their energy out directly, discarding the rest. Well, as it turns out that's not such a good thing, for a couple of reasons. One is that we have to keep mining or drilling or harvesting to feed an insatiable need for more fuel. The other is that what we throw away isn't exactly environmentally benign. Hence the pickle we're in with global warming and air pollution. On top of all that the constant drilling and mining, besides being environmentally insulting, is a catalyst for wars or, at the very least, economic strife.

The metal-fueled car, however, wouldn't require constant sourcing of new metal aside from the amount needed to keep up with growing demand. The fuel would take its energy from the heat and the gases used to separate the oxygen from the metal oxide after it's been combined in powering the car. Whatever is the source of that heat and those gases is actually the primary energy source. As you can easily imagine, the primary energy source proposed in this book is fast reactors. Use that indirectly to drive a metal-fueled car and you're essentially driving a nuclear-powered automobile.

While the researchers at Oak Ridge are predictably enthusiastic about burning steel in cars, others aren't so excited. One of the problems is weight. Steel is mighty heavy, and the rust that would have to be carried around would be even heavier. Though the rust would be swapped out when refueling, that means there would be transportation costs two ways instead of

one way as with gasoline, plus the full weight of the fuel — or
more — is always on board. On the other hand, steel and rust
can be carried around in regular trucks instead of tanker trucks
and both are safe to transport, so that knocks the trucking costs
down a bit. But what about the weight?

Not a problem, say the Oak Ridge boys. Steel isn't the only
metal that can burn. Aluminum will yield up to four times the
energy per pound, and boron up to six times the energy. But
aluminum costs about fifteen times as much as steel, and boron's pretty spendy too. Here's where the hydrogen guys step in
again crowing about their pet fuel, because of course in terms
of energy per pound hydrogen weighs nearly nothing compared
to the energy it will deliver. The problem, of course, is that it's
devilishly hard to store and move around.

Here's a graph[136] to help visualize the sort of energy factors
we're talking about, showing the energy per unit mass on the

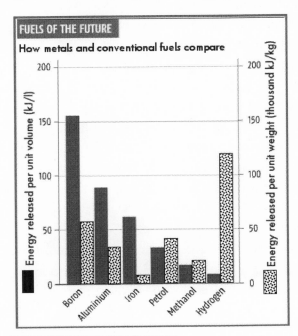

---

[136] Kurt Kleiner, "Powdered Metal: The Fuel of the Future," *New Scientist*
Oct 22, 2005.

right side of each pair of bars and energy per unit volume on the other. We can clearly see both the upside and the downside of the hydrogen story there. Methanol's looking pretty iffy too. Boron — the fifth element — is clearly superior to all the others, providing far more energy per liter than any of them, and much more per kilo than all but hydrogen. So why do the Oak Ridge researchers seem to be ignoring boron in favor of steel? If it's just price, should that really be an issue if it can be 100% recycled?

Therein lies the rub. Boron just won't burn in air. The darn stuff doesn't want to light. At the nano scale that the Oak Ridge researchers are working with there might be a better chance, but the necessary venting of the exhaust would contaminate and lose some of the original boron. That would be a problem with burning steel or aluminum in air, too, for one would have to expel all the other gases in the air (mainly nitrogen, some argon, and traces of other elements). There are bound to be some minute metal particles that depart with the exhaust, as well as oxides of nitrogen, which are pollutants.

So burning metal in air — if it can be persuaded to light — thus entails at least some degree of metal loss and a resultant need for continuously replenishing the supply from outside sources — mining, ultimately. Nor is metal burning anywhere close to pollution-free. If boron (which actually isn't a metal, per se) could be made to burn in air by virtue of powdering it exceedingly fine, it's too expensive to discard even small amounts. Plus, of course, we don't want to pollute. Zero emission is the goal.

Enter a creative Canadian from a small town on the shore of Lake Ontario, Graham Cowan.[137] He has been pondering this for nearly a decade and early on came up with the inspiration of burning boron in pure oxygen. Therein lies the key. For not

---

[137] Graham Cowan, *Boron: A Better Energy Carrier Than Hydrogen?* (2007 [cited]; available from http://www.eagle.ca/~gcowan/boron_blast.html.

only will it burn — very hot! — in oxygen, but you won't need an exhaust pipe. Inject the oxygen under pressure into a turbine or heat exchanger with boron and the result is boron oxide, which at high temperatures is a syrupy substance but which, once it cools, forms a glassy and quite non-flammable ingot. Any unburned oxygen or boron can simply be rerouted back to the intake again. Thus all the boron would eventually be captured as boron oxide, as would all the oxygen. Here's another bonus: Hydrogen, or typical fuel hydrocarbons like gasoline or diesel, require almost twice as much oxygen as boron per unit of energy produced.[138]

Precisely what form the boron should take for easiest handling and most efficient burning remains to be seen. As a hard solid substance, boron could be pelletized, or formed into long bands and wound onto spools, or perhaps powdered like the Oak Ridge researchers envision for their steel-burning engines. When a boron car driver went to refuel, she'd simply drop off the boron oxide for recycling and refill her tank (or spool) with new boron, receiving a credit for the oxide. Boron's not cheap. The amount equivalent to a 20-gallon tank of gasoline would cost somewhere in the range of a couple hundred dollars. It can be mined (as is currently done most famously in Boron, California) from the area of ancient lakebeds, or it can be extracted from seawater. There is plenty of it in the world, especially since it would only have to be extracted once for each vehicle.

Here's where that two hundred dollars becomes vanishingly cheap. The boron oxide would be hauled back to a recycling center. There it would be heated to about 700° Celsius and processed with a couple of catalysts to drive off the oxygen, which could then be released into the air (or put to any other use). The catalysts would be retrieved for reuse and the now-pure boron reformed for shipment back to the fueling point — every bit of it.

---

[138] Ibid. ([cited).

There are a number of marvelous aspects to this system. For one, boron is completely safe to carry around, so it can be transported by truck or train at the cheapest rates available. It's dry, odorless, and virtually inert, so it can be stored indefinitely almost anywhere: in a garage, in a basement, or in the backyard. Yet this is only one of boron's many spectacular advantages compared to hydrogen or nearly any other new fuel, for this system completely eliminates the chicken and egg infrastructure problem that can cripple the introduction of new technologies.

If you were the first person in Alaska to own a boron car and there was only one boron recycling plant in the country, in Florida, it would still be perfectly feasible to drive a zero-emission boron car economically. Just have a few totes of boron trucked up from Florida and store it in your backyard. Refueling would, unfortunately, be restricted to home, but for a long trip you could just put a bunch of boron in the trunk. For a trip all the way down the Alcan Highway — to go visit the boron supplier, perhaps — you could just fill your back seat with boron to boot, or pack a cartop carrier full of it, or even put a bunch in a trailer. But for the most likely usage patterns, home as your sole fueling station would work fine. When the boron started to run low, you'd just ship the collected boron oxide back to Florida and order some new fuel.

Of course you wouldn't be the only guy in Alaska to have a boron car for long, so fueling stations would soon be popping up all over. The investment would be about nil. The 7-11 store on the corner could sell boron. No need for underground storage tanks, hazardous materials permits, or fire suppression equipment. Rather than requiring a huge infrastructure investment to make boron practicable as a fuel, all it would take is the investment in a single boron recycling center.

Those recycling plants would be powered by IFRs. The temperature of the sodium circulating in an IFR reactor is almost hot enough to drive off the oxygen from the boron oxide.

A bit of electrolysis is needed later in the process to recover the magnesium and chlorine catalysts, which are 100% reusable, and of course the IFR can also produce the electricity needed for that. By using the thermal energy straight off the reactor for preheating the boron oxide, the efficiency of the recycling process would be improved over using electrical heating for the entire process. The small temperature shortfall can be rectified with a heat boost provided electrically. Remember, the fuel for both the heat and electricity in an IFR is free.[139]

So how much would a tank (or spool, or bin) of boron cost? Well, as mentioned previously, to extract enough from land or sea for a tank of fuel would cost a couple hundred dollars. It takes about a pound and a half of boron to equal the energy of a gallon (U.S.) of gasoline, though of course how efficiently that would be converted to power would depend on the engine design. For purposes of familiarity, from here on till the end of this chapter I'll just talk about 1.5 pounds of boron as a gallon, meaning that it's equivalent in energy to a gallon of gas, though it would weigh about a fourth as much and take up less than a quarter the volume.

Once you've bought your new boron car and paid that couple hundred dollars for its first tank of boron, you'd never buy any really "new" boron again. The only costs would be the recycling, and since the IFR fuel that would power the recycling process is essentially free, that processing charge would be minimal. Then you've got shipping costs of the boron from the plant to the store on the corner, and shipping the boron oxide back to the recycling plant.

---

[139] Throughout this book, when I write that the IFR fuel is free that refers to the fuel alone, to which must be added the usual cost of the metal parts of the fuel assemblies (cladding, etc.). When spent thermal fuel is being used up in IFRs, there will also be the fixed cost of reprocessing it into IFR fuel. Once IFRs are all running solely on depleted uranium, all reprocessing will be done on-site at the IFR and will be part of normal operational costs. The fuel itself will, indeed, remain free.

After it burns in oxygen the weight of the boron oxide is about three times that of the original boron. So the weight and volume advantages that looked so good just a couple of paragraphs ago compared to gasoline turn out to be just about a wash. Nevertheless, in no time at all there would be boron recycling plants springing up in every state, and because boron (and boron oxide) can be shipped as cheaply as gravel because of its safety, shipment to and from the recycling centers will cost on the order of about two cents a pound each way, or about 12¢ per "gallon".[140] As for the cost of the recycling itself, the price tag to build the plant is of course a major capital expense, as it will be for all the IFR facilities. The catalysts magnesium and chlorine are part of that capital cost because they're 100% recoverable and reusable. Only the operating costs, and the amortization of the plant cost, are involved.

I will grant the reader, at this point, that I can only guess as to the ultimate cost of the recycling, since in order to figure out the amortization costs I would have to know the amount of fuel that such a recycling plant could process over its expected life span of many decades. I'll get more into the costs of building IFR plants in a later chapter, but I believe it would be safe to say that the cost of recycling a "gallon" of boron would be negligible, probably pennies. Let's say eight cents. Added to the cost of shipping you'd be looking at twenty cents to the store. The storekeeper takes a nickel; you've got your boron fuel for two bits a "gallon." If you want to quibble with my back-of-the napkin calculations here, feel free to do so. But clearly this fuel will be staggeringly cheap compared to anything you've used since the Sixties.

As for the cost of the recycling plants, even if your tax dollars pay for them it will be a bargain, considering the safety,

---

[140] Remember, a "gallon" of boron weighs only 1.5 pounds, but the resulting boron oxide weighs three times that much, for a total of 6 pounds that must be shipped back or forth at 2¢/lb.

economy, and — last but definitely not least — the zero emissions. Not only will we be nipping global warming in the bud, but the heretofore polluted air in our cities will soon be astoundingly clean. As energy use becomes converted wholesale to IFR-produced electricity and the cars to boron, city air will become as refreshing as country air. From the upper floors of the buildings in downtown Los Angeles you'll have a grand view of the mountains to your north and east, way across San Bernardino. Walk into the Zocalo in Mexico City and you'll once again see the windows on the palace across the block. It will be a far different, and much more enjoyable (and healthier!) world.

But how about this pure oxygen business? Is it even possible to extract pure oxygen from the air in the quantities needed with an oxygen extractor that will fit in a car? In theory, definitely, but there hasn't been much of a reason for people who need pure oxygen to work on miniaturization of the equipment. There are technologies capable of supplying pure oxygen, such as Nafion or zirconium oxide, which fuel cell researchers have been working with. A NASA researcher who's worked with oxygen extraction technologies for aircraft also told me about a system from which the oxygen exits at high pressure (which we want for the car) and very hot, about 2,000°C. Injecting it at that temperature into the engine would be like a preignition system. We want as much heat as possible, after all.

The issue of how much space we'll need for oxygen extractors is mitigated to a great degree by the lack of volatility of boron itself. Unlike gasoline-powered vehicles, there need be no shielded area in which to carry the fuel (or the boron oxide). Likewise the oxygen extraction equipment could be placed pretty much anywhere. You could even carry boron inside bumpers or quarter panels. And bear in mind that the oxygen is utilized 100% and that boron combustion only sips oxygen compared to other fuels.

Now we have to convert the heat into mechanical energy for the car, and supply the initial power to start the whole process. For this and a reason I'll explain presently, all the boron cars would be boron/electric hybrids. The batteries wouldn't have to be nearly as large as current hybrids, because they'd only have to provide power for the beginning of the trip while the oxygen extractor kicked in. As for the actual fuel ignition, there are various options.

Graham Cowan has worked on this aspect of the idea quite a bit, figuring out the potential problems with the exceedingly high temperatures in a boron turbine, the laminar flow of the hot syrupy boron oxide along the blades, the type of materials necessary, etc. Brittleness of the turbine blades is definitely an issue when you're talking about materials that can stand that kind of heat over time. While I would defer to Cowan's knowledge and inspiration in these areas over my amateur speculations, should the direct turbine approach prove overly difficult, a simple combustion chamber with a water jacket (or perhaps another liquid for moving the heat to a turbine) would likely work just fine. It will take some R&D to figure out the best configurations to extract sufficient heat to run the car, but considering the prodigious heat that boron puts out — and let's not forget the hot oxygen you're putting in there — it's a certainty that a steam turbine could easily run off it. Boron burns considerably hotter than hydrocarbon fuels, so there's quite a cushion there to compensate for a possible drop in overall efficiency. The internal combustion engines we use today aren't exactly models of efficiency anyway, converting only about 15% of the energy in their fuel to the intended purposes.[141]

As for transferring the turbine power to the drive train, there are compelling reasons to simply run a high-powered

---

[141] fueleconomy.gov, *Advanced Technologies & Energy Efficiency* (US DOE/ EPA, 2007 [cited]); available from http://www.fueleconomy.gov/FEG/atv. shtml.

electrical generator and drive the wheels with electric motors. The reason for that has to do with the rest of the energy systems that people in the post-revolution era will be using.

Bear in mind that once all the coal plants are replaced with IFRs, the power plants burning natural gas will be the next to go, and home heating will also be converting over from natural gas and heating oil to electric heat, most probably with heat pumps to provide both heating and cooling with greater efficiency than resistance heating systems and separate air conditioners. While not the focus of this book, I would mention in passing that geoexchange heat pumps are wondrously efficient and it would constitute wise energy policy to subsidize their installation (perhaps from electricity revenues), both in new homes and retrofits. Retrofitting them is hardly more difficult than new construction, consisting mainly of digging up a section of your yard to lay out the heat exchangers, then patching your heat pump into your existing furnace and A/C system. Quiet, amazingly cheap to run, almost maintenance-free, and zero emissions (remember, the electricity to run them is coming from a zero-emission IFR plant).

Okay, so you've got a nice all-electric home and you're living in Winnipeg in the middle of winter and suddenly the power goes out. Not a problem, because your house has a fat cord in a little utility box on the side closest to your driveway, and your boron car has a plug on it that can feed power from the car's robust alternator. Put on your hat and mittens, grab your parka and the car keys, run outside, plug your house into your car, and start it up. Then hustle back into the house. Oh, but you forgot to fuel up today! Don't sweat it. Because of boron's safety and stability, every homeowner in severely cold climates would keep an emergency supply in the garage, closet, barn, or basement. Pretty nice to have that portable electrical plant when you take those summer trips to your off-the-grid cabin in the summertime, too. Honda won't be too happy when their portable generator sales drop off to nothing, though. But hey, that's evolution.

Besides, Honda and all the other carmakers will be too busy to notice. They'll be in the heyday of a new automotive Golden Age, replacing an entire planet's fleet of vehicles. And who won't want a boron car? The fuel savings alone will sell them, never mind the zero emissions. Of course as boron takes over, gasoline prices will plummet, especially once OPEC becomes nothing but a distant and quite unpleasant memory. We'll explore those ramifications of the revolution a bit later.

For now, just sit back and picture how this all fits together. All over the world IFR plants, assured of hundreds of years of free fuel, are silently humming away, supplying power for not only the old uses but for steadily evolving industrial applications as well. No more coke smelters for steel production. They, like all the other coal and natural gas users, have switched to electricity. Homes and business are all electric. Even the busiest intersections on the most sweltering days have nary a hint of exhaust smell, the air is clear and fresh. Your kids grow up knowing blue skies and distant horizons even in the biggest cities, their only experience with smog being from history class. And at night those cities could be spectacular, with skyscrapers outlined in lights (LEDs or CCLs, of course).

I freely grant that there are R&D challenges ahead for the boron car, but they are most certainly surmountable in the near term. The most difficult part of it will likely be the size of the oxygen extractors, but if that took too long we could still initiate the boron engines and carry oxygen tanks on board, which we could fill up every night at home from a small extractor/compressor system outside the garage. On long trips they could be swapped out with standardized tanks at fueling stations. Given the great deal of latitude afforded us by the theoretical limits of oxygen extraction, though, it's highly doubtful that would be necessary. Already five years ago oxygen extractors were almost small enough, even with their efficiency being barely 5% of the theoretical limit. Give that challenge to the wizards at Sandia Labs and sit back and watch the fur fly. We'll be tooling around in borocars in a heartbeat.

On the other hand, there are a lot of new electric car technologies on the horizon that seem to show great promise, from the aforementioned Phoenix to high-tech capacitor systems. And work being done on so-called flow batteries holds out the possibility of being able to simply pump out discharged electrolytes and pump in a fully charged solution, which wouldn't take much longer than fueling up with gasoline today. It's possible that by the time this book is in your hands a viable electric car will be on the road. What use for boron then? Well, you still have that home in Winnipeg, remember? And long trips in remote areas could be impossible with all-electric vehicles, though for most uses they would be just peachy. The average car trip in America is about 29 miles, so usually it would work just fine to plug in at home. If Phoenix Motorcars actually succeeds in building a car with long range per charge and a ten minute charge cycle as they're promising, admittedly the need for boron will be minimized. Nevertheless it could well be used in trucks, trains, heavy equipment, portable generators, or for safely and cheaply transporting energy in areas (such as much of the developing world) where power grids are inadequate or nonexistent. Our Winnipeg family could get by just fine with an electric car, though, as long as they kept a boron-powered generator out in the garage.

A boron/electric hybrid, however, would be the best of both worlds. Not only would you have terrific range even beyond the grid, but the charging cable that plugs into your house every night (assuming we make these plug-in hybrids) could operate in reverse if the power went out. All you'd have to do is start the car to kick in the boron power. Of course with a truly efficient boron/electric hybrid you might drive around with a tank of boron for months before ever having occasion to use it.

Would that be a bad thing? Absolutely not. From an efficiency standpoint it would be the best situation. Any time energy is converted from one form to another it incurs an energy penalty. So it would be more efficient just to use electricity straight from the IFR to charge up our cars. Dependable boron/electric hybrids would mean only that we'd need fewer boron recycling plants, saving both money (especially the high capital cost) and energy.

Yes, I am a technology optimist. But look at the challenges to boron car development compared to any other alternative energy technology. While not inconsequential, they are certainly surmountable in the near term, and a boron system virtually eliminates the chicken and egg infrastructure problem. A single recycling plant built anywhere in the country would enable boron cars to take to the road nationwide.

Graham Cowan once took a walk like Daniel into the lions' den. He presented a paper[142] at a convention of hydrogen researchers, explaining the superiority of boron as an energy carrier. Lots of stunned faces, but no fatal flaws were even suggested. Couple boron cars with IFR deployment and you've got yourself a brave new energy world.

---

[142] Graham Cowan, "Boron: A Better Energy Carrier Than Hydrogen?," in *11th CHC Hydrogen Research Conference* (2002).

CHAPTER SIX

A Decidedly Immodest Proposal

Always listen to experts. They'll tell you what
can't be done and why. Then do it.
— Robert Heinlein

T HERE IS PERHAPS no field of scientific endeavor more
rife with misinformation, ignorance, passion and hysteria
than the field of nuclear power. Sorting through it all with
the threat of global warming looming on the horizon is akin
to being diagnosed with cancer, and forced to make informed
life-or-death decisions quickly, wading through volumes of
purported cures, the vast majority of them utter quackery.

It's no wonder politicians seem to have an incoherent position on nuclear power, since depending on whom they listen to
they might either believe it's terrific or apocalyptic. It doesn't
help that professionals of one stripe or another can be found on
both sides of the debate, or that a lack of knowledge or perspi-
cacity often is considered no barrier to making policy recommendations. Such is the case of a 2003 MIT study called "The
Future of Nuclear Power." Its impressive array of professionals
suggested the establishment of a "large nuclear system analysis,
modeling, and simulation project...to assess alternative nuclear
fuel cycle deployments..." For this study they anticipated years

would be required. Until that could be completed, they recommended that the nuclear power industry "halt development and demonstration of advanced fuel cycles or reactors until the results of the nuclear system analysis project are available."[143]

This study was completed some nine years after Congress prematurely shut down the IFR project at Argonne National Laboratory (more on that later) just as they were ready to demonstrate the last step of their closed fuel cycle program proving the commercial viability of pyroprocessing. The equipment was in place, the trained personnel were there to run it, and at the last minute Congress balked and jerked the funding rug right out from under them. Nine years later the MIT group suggested several years for more study — and, moreover, a halt in development — despite the fact that nearly a decade had passed since the commercial-scale fuel cycle could have been amply demonstrated within a year or two. If this reminds you of the tired and irresponsible argument that global warming is just a theory and we have to study it more before we consider doing anything, you will be forgiven. But forgiveness is much harder to come by for the august panel making these recommendations, especially since just a bit earlier in the paper they freely admitted, "We know little about the safety of the overall fuel cycle, beyond reactor operation."[144] Notably absent from the panel, by the way, were any of the physicists or engineers who worked on the Argonne project — the most qualified professionals in the field who'd led the work at the nation's center for nuclear power research since WWII — though there did seem to be ample room for some politicians. I suppose the Argonne people were all too busy *not working anymore.*

With advice like that, it's little wonder that the U.S. government is putting off the decision as to whether to close its

---

[143] Eric S. Beckjord Exec Director, "The Future of Nuclear Power," (Massachusetts Institute of Technology, 2003).

[144] Ibid.

nuclear fuel cycle until the year 2030. You will recall that a closed fuel cycle refers to a system that burns all the actinides and results in a much smaller quantity of nuclear waste (as little as 1/10 the amount that the current generation of reactors produces) with none of the long-term waste disposal issues. The MIT panel's recommendation? Build more nuclear plants with current designs and one-pass fuel cycles, continuing the buildup of nuclear waste instead of diminishing it with the deployment of fast reactors. Do we really have to wait another 23 years to figure out that we want less nuclear waste instead of more?

Apparently more rational voices have been heard, however, because in 2006 the United States began working with several other countries on the Global Nuclear Energy Partnership (GNEP), which focuses on development and deployment of fast reactors and pyroprocessing facilities, among other things. GNEP would take up where Argonne left off when their nearly complete program was prematurely terminated. It will demonstrate the commercial application of the entire IFR/pyroprocessing concept, the closed fuel cycle, which addresses both proliferation and nuclear waste concerns. The downside of GNEP, in my humble opinion, is that it casts too wide a net, concerning itself with a variety of nuclear power research projects. Given the political tenuousness of all things nuclear, I fear that such a melange of objectives may well doom the entire project. Far better, it would seem, to focus on the revitalization of the IFR and let other projects be dealt with separately.

Full Disclosure

At this point I feel I should clear something up. I am not, nor have I ever been, associated in any way with the nuclear power industry. Nor do I have any connection whatsoever, financially or otherwise, with any energy industries or research organizations. Aside from a number of mostly retired nuclear physicists, the industry doesn't even know I exist. A little further on in this book you'll understand why I prefer that to be the case.

I have always considered myself an environmentalist, and still do — more than most, actually, since I've devoted years trying to figure out how our planet can be kept livable. I am also a pragmatist. As much as I would like to believe that solar power, wind power, biofuels, curly light bulbs and bike riding will solve our global warming and other energy dilemmas, it is abundantly clear that it will take much, much more than even their most zealous promoters can hope for in their wildest energy utopia dreams. World energy use is projected to double by mid-century, though some say it will only take half that long[145]. We need large amounts of baseline energy and, unless you're a global warming skeptic, you must agree that we can't afford to wait around for wondrous new technological breakthroughs. Heck, even if you *are* a global warming skeptic, you still must like the idea of clean air, abundant power, cheap fuel for your car, and no more oil wars.

Okay, with that out of the way I'd like to get back to that MIT study. While recommending the building of new nuclear power plants (provided their retro criteria can be met), the study projects 1,000–1,500 nuclear plants to be operating worldwide by the year 2050. This would help ameliorate the effects of global warming somewhat, they contend. Yet whereas today about 17% of electrical generation worldwide is provided by nuclear power with nearly 450 plants online, a thousand plants in 2050 would be providing just 19% (because of the ever-increasing demand for electricity). That means that non-nuclear power plants — mostly coal

---

[145] Keith Bradsher, "Emissions by China Accelerate Rapidly," *International Herald Tribune* Nov 7, 2006.

and natural gas, presumably — would be churning out *way more* global warming gases than today. At the rate the MIT panel envisions things, we would actually be falling far behind when it comes to climate change, and we'd be 42 years farther down the road.

Lest I belabor the point too much, I will go just a step further in commenting on the MIT study because it is symptomatic of the sort of disinformation that confuses the public and hinders the development of enlightened policy, despite issuing from "experts." In arguing for continued deployment of thermal reactors, they contend that the only disadvantage of those designs is the issue of nuclear waste. As for the IFR plants, the panel alleges disadvantages in cost, short-term waste issues, proliferation risk, and fuel cycle safety.

Ironically, it is precisely all these issues that the EBR-II project at Argonne was designed to solve, challenges they overcame with spectacular success. The safety and viability of the closed fuel cycle was one of their foremost priorities. The pyroprocessing system that was on the brink of full demonstration mode is a system of electrorefining currently used in many industries, a relatively simple process. When the project was closed down it was not because of any concern with the safety of going ahead with the demonstration. It was, as you will see in a later chapter, purely political in nature. It's baffling that this study cautions about short-term waste issues, since the IFR is designed to burn nuclear waste from thermal plants and produce a fraction of that amount in a considerably safer form, for disposal without long-term radioactivity concerns. Not only that, but in the short term there would be no waste at all issuing from IFRs. They produce so little waste that it could easily be kept on-site for the entire lifetime of the plant, then all safely moved out at once when the plant is decommissioned.

It's a mystery where the MIT group came by their implied assumption that IFR plants would be more expensive than thermal plants. After all, the system is considerably more straight-forward than the LWRs currently in use, and which they're recommending as the ones we should keep building. The use of unpressurized coolant and heat transfer systems and the passive safety features of the IFR mean that there will be significantly fewer valves, pumps, and backup safety systems needed. As for the pyroprocessing that forms the "integral" part of "Integral Fast Reactor," it would readily lend itself to modular construction and when mass-produced should actually be quite cost effective, as we'll discuss in greater detail when we explore the economics of IFRs.

Of all the misconceptions and misrepresentations that are evident not only in the MIT study but in anti-IFR objections from people who should know better, proliferation risk is the most unjust and wrongheaded argument against it. Yes, fast reactors can be used to breed plutonium. As we've seen, all nuclear reactors do. The operation of IFRs is actually far safer, from a proliferation standpoint, than the thermal reactors that the MIT panel is recommending. It would, in fact, be considerably easier for a would-be member of the nuclear club to irradiate some fuel rods in a thermal reactor for only a short period, and use the PUREX process to obtain weapons-grade fuel—especially if they already had a PUREX plant— which is in fact what North Korea has recently done.[146] Diversion and reprocessing of fuel from an IFR would be substantially more problematic and easily detected. One of the Argonne physicists laid out the facts:

> Expert bomb designers at Livermore National Laboratory looked at the problem in detail, and concluded that plutonium-bearing material taken from anywhere in

---

[146] The North Korean reactor was a graphite-moderated reactor, not an LWR. The spent fuel, however, has the same characteristics and reprocessing potential as LWR fuel.

the IFR cycle was so ornery, because of inherent heat, radioactivity and spontaneous neutrons, that making a bomb with it without chemical separation of the plutonium would be essentially impossible — far, far harder than using today's reactor-grade plutonium.

First of all, they would need a PUREX-type plant — something that does not exist in the IFR cycle.

Second, the input material is so fiendishly radioactive that the processing facility would have to be more elaborate than any PUREX plant now in existence. The operations would have to be done entirely by remote control, behind heavy shielding, or the operators would die before getting the job done. The installation would cost millions, and would be very hard to conceal.

Third, a routine safeguards regime would readily spot any such modification to an IFR plant, or diversion of highly radioactive material beyond the plant.

Fourth, of all the ways there are to get plutonium — of any isotopic quality — this is probably the all-time, hands-down hardest.[147]

Even if diversion of weapons-grade material wasn't so improbable, it's not even really an issue if you're talking about massive deployment of IFRs to combat global warming. Just look at the countries that, now and in the next fifty years, will be producing the vast majority of humankind's greenhouse gases (GHGs). Seventy percent of GHGs are produced by countries that already have nuclear weapons — with the exception of Japan, which has full reprocessing capabilities anyway.[148] Given the rate at which India and especially China (both nuclear club members) are bringing new dirty

[147] Stanford, "Integral Fast Reactors: Source of Safe, Abundant, Non-Polluting Power."

[148] Lenntech, "The Global Warming and the Greenhouse Effect."

coal plants online, a couple of decades hence we'll likely be talking about the producers of upwards of 80% of GHGs. These countries alone should embark on a crash program of modern nuclear plant deployment to completely replace all their electrical generation with state-of-the-art passive safety thermal plants and, as soon as possible, with IFR technology. The decades that it would take to effect the conversion would leave plenty of time to figure out secure arrangements for the other nations of the world to employ the technology, without having to be concerned about proliferation.

As this technology is deployed, a lot of non-nuclear nations will want to avail themselves of it too. One way to do that would be to create an international energy consortium that would have trained plant operators running all the plants in at least the non-weapons states. The plants themselves could be operated as international energy zones with the same independent jurisdiction as embassies now enjoy. Teams of operators would be trained and managed by the consortium and transferred randomly from one country to another to prevent collaboration with shady governments that might wish to attempt a diversion of fuel in collusion with plant operators. Unlikely, perhaps, but there's no reason the system couldn't be set up that way just to err on the side of caution. Ironically, considering the misguided fears of proliferation with the IFRs, if such a diversion were to be attempted it would probably happen with the incoming nuclear waste from thermal reactors. But any sort of skullduggery like this would be highly improbable since the international operation would assure complete control at all times with random inspections by the consortium's oversight group.

We should really go beyond confining the consortium's power plant operators to non-nuclear club countries, though. Besides merely providing operators, the group would also assure universal safety standards and inspections. Leaving such an inspection regime to individual nations would be asking for trouble, as we've already seen too often with cozy relationships between power plant operators and regulatory agencies. Far better to assign custody of all nuclear power plants to the international consortium, no matter where the plants are located. Countries with weapons programs already in hand have separate facilities for those purposes anyway. There is no compelling reason to undermine international standards of nuclear power plant safety for the sake of outmoded notions of sovereignty, or whatever other excuses might be trotted out.

There is another technology just on the horizon that perhaps will be even better for politically uncertain nations that want nuclear power. This is a sealed system that has been coined a "nuclear battery." The Toshiba Corporation is waiting for permission from the U.S. Nuclear Regulatory Commission to deploy the first of its kind in a small town in northern Alaska called Galena. The 10-megawatt power plant would be buried in a concrete silo with only two pipes rising up out of the ground to supply steam to a turbine generator on the surface and to feed water back down. The reactor itself would not require any operator intervention and would provide power for 15-30 years before it needed to be recharged. At that point it could be unsealed and the fuel assembly could be extracted and replaced for continued operation.

These nuclear batteries can be made in various sizes, from quite small as in the Galena case, to large enough for providing power to cities. They represent a form of distributed generation that is especially attractive to developing nations lacking existing power grids. The possibilities for Third World nations in desperate need of electricity are stunning. The reactor technology is essentially the same as the IFR, only it's smaller and sealed for essentially automatic operation. The same passive safety is built in, and any actinides unused when a battery is finally refueled could be reprocessed at a full-fledged IFR and burned again.

The realistic options available to humankind for quickly and effectively addressing the urgent situation in which we find ourselves are distressingly few. Denial is not one of them. Putting on rose-colored glasses and dreaming of a happy world of spinning windmills and vast seas of solar panels isn't either. We need a whole lot of clean energy and we need it now. Not only that, but we'll be needing plenty more going forward. Given the welcome albeit insufficient contributions of wind and solar and whatever other truly clean energy systems will be added to the mix, the gigantic shortfall can be met by two options: fossil fuels or nuclear power. At best, the unproven technology of carbon sequestration could allow us to continue using some fossil fuels, though all the coal plants on the drawing board—even in the USA—won't employ it. Even if they did, they would still have to come up with an environmentally benign disposal method for all their ash, laden with heavy metals and radioactive elements. Those highly polluting plants have expected life spans of about sixty years. If we're really serious about global warming we will need to confront the necessity of shutting down perfectly serviceable and even relatively new power plants. Building even more of them would be foolhardy.

That leaves nuclear power. So the question becomes: do we want nuclear power systems that add to an ever-growing pile of eternally (for all intents and purposes) radioactive nuclear waste, lack passive safety systems, and stand to run out of easily available uranium as their numbers increase? Or do we want power plants that clean up the nuclear waste from the older reactors, eliminate weapons-grade materials from the fuel cycle, are meltdown-proof, and have enough free fuel already on hand for nearly a thousand years?

The phony debate that might be inferred from all the misinformation and outright lying about IFRs is like many of the other faux controversies that seem to plague us these days. Ignorance of the facts can only be used as an excuse just so much, and policymakers and scientific advisors should not be indulged. All too frequently, as in the MIT study, one piece of the story will be accurate while ten more will stand in direct opposition to it, so veracity ends up lost in the shuffle. Anyone who's been attentive to politics and the media the past several years will recognize the style. Read what the people who actually worked on the Argonne IFR project have to say,[149] then judge for yourself.

In all fairness to at least one of the scientists on that MIT study, though, I would like to digress for a moment. It is said that a camel is a horse put together by a committee. If that is an apt analogy, then a horse put together by a committee with politicians on it might turn out to be an ostrich — all the more so because of how so many politicians are more than willing to bury their heads in the sand. We have seen many scientific/ political collaborations end up with watered-down conclusions with which the scientists on the studies are outraged, a pertinent example being some of the recent IPCC studies on climate change.

Thus the final MIT report on their nuclear energy study ends up characteristically cautious, its recommendations wholly inadequate to deal realistically with the overwhelming seriousness of global climate change. Yet at least one of their number seems quite aware of the potential of IFR technology. He is the current director of Woods Hole Research Center, Dr. John Holdren, a physicist with a formidable intellect and impressively eclectic experience. At an energy symposium at Berkeley in 2006,[150] Dr. Holdren presented statistics, compiled and reconciled from a wide variety of sources, laying out the best estimates of ultimately recoverable global energy resources.

---

[149] http://www.sustainablenuclear.org/PADs/pad0509till.html
http://units.aps.org/units/fps/newsletters/2006/january/article2.cfm
http://www.nationalcenter.org/NPA378.html
[150] John P. Holdren, *Global Energy Challenges & the Role of Increased Energy Efficiency in Addressing Them* (Berkeley, California: 2006).

Since the IFR clearly has the potential to provide all the energy humanity needs provided we deploy it in sufficiently large numbers, and since Dr. Holdren's statistics included the use of uranium in breeder reactors, I decided to compare the relative abundance of various resources. To do this I first did a rough calculation of the amount of energy that humanity can be expected to need over the next fifty years.

Though world energy demand is predicted to double by 2050, I strongly suspect it will at least triple and possibly even quadruple. Here's why: For starters, it nearly doubled in the last 25 years. With the developing countries having so far to go, and with some of them galloping ahead at breakneck speed, doubling again within the next 25 seems entirely plausible, especially if they're offered the wherewithal to do so. So at least tripling demand by mid-century seems like a shoo-in. Now here's the kicker: By 2050 or shortly thereafter it's expected that the world's population will increase to about ten billion, according to most projections. Many countries are already having serious water supply problems, so massive desalination and canal projects are going to absolutely demand a huge amount of energy. That doesn't seem to have been factored in to most projections, or at best only perfunctorily. Such an oversight is, I believe, naive in the extreme. The only thing that can obviate the need for such massive projects is death on a massive scale, from whatever source: pandemic (intentional or unintentional), war, famine, or a combination thereof. Assuming that we intend to avoid such global calamities as much as possible, we have to plan for future water demand, and that means being realistic about energy demand for freshwater production and distribution.

That being said, I would suggest that a tripling of energy demand by 2050 is entirely reasonable to expect. It never hurts to plan for a worst-case scenario if we can. Since humankind used 16 terawatt-years (TWy) of energy in 2005, and will likely use triple that in 2050, we should be able to calculate an average use from now till then of about 32 TWy/year (double the current usage). If we assume that population will begin leveling off by then, or that energy use will be more efficient (two optimistic assumptions not entirely fanciful but hardly certain), we can divide the numbers in Dr. Holdren's statistics by 32 to get some idea of the number of years of energy available from each of these sources, and graph them accordingly, assuming that any one of them would be providing all the world's energy. (Renewables like wind, solar, and hydro obviously would somewhat lessen the demand, but for purposes of this comparison we'll show how long each single source could last if it was called upon for the entire planetary energy demands we can reasonably expect to confront.) This is obviously not what would happen with oil, gas, or even coal, but since it is entirely possible to provide for all those energy needs solely with IFRs (should we choose to do so) we will compare apples to apples here.

Dr. Holdren categorized his energy sources thusly:

- Conventional oil: ordinary oil drilling and extraction as practiced today
- Conventional gas: likewise
- Unconventional oil (excluding low-grade oil shale). More expensive methods of recovering oil from more problematic types of deposits
- Unconventional gas (excluding clathrates and geopressure gas): As with unconventional oil, this encompasses more costly extraction techniques
- Coal: extracted with techniques in use today. The worldwide coal estimates, however, are open to question and may, in fact, be considerably less than they are ordinarily presented to be.[151]
- Methane Clathrates & Geopressured Gas: These are meth-

---

[151] David Strahan, "The Great Coal Hole," *New Scientist*, Jan 19, 2008.

ane resources that are both problematic and expensive to recover, with the extraction technology for clathrates only in the experimental stage.

- Low-grade oil shale: Very expensive to extract and horrendously destructive of the environment. So energy-intensive that there have been proposals to site nuclear power plants in the oil shale and tar sands areas to provide the energy for extraction!
- Uranium in fast breeder reactors (IFRs being the type recommended by the author)

The resulting graph was essentially useless in the normal two-dimensional bar graph format, since in order to accommodate the bar representing the number of years that the entire earth could operate using IFRs alone, most of the other bars would be invisible. Here is the graph, with the numbers below indicating how many years each type of energy source would last if we used it alone for all the world's energy needs. The bars are shown from above so you can see that there's at least something there. You may note that the number of years of economically recoverable uranium for use in IFRs is a thousand times greater than if we'd use it in LWRs. Since LWRs use about 1% of the energy available in uranium ore, that would seem to indicate an exaggeration of IFR fuel availability by a factor of ten. But it's not a mistake. Since uranium is utilized so much more efficiently in IFRs, it would still be quite economical to extract that much more uranium from even low-grade ores than if it was only going to be used in LWRs. In fact, uranium can be extracted from seawater, and certainly that technology will be perfected long before 50,000 years have passed. It is, in essence, an unlimited resource, which is why it can legitimately be compared with renewable energy systems.

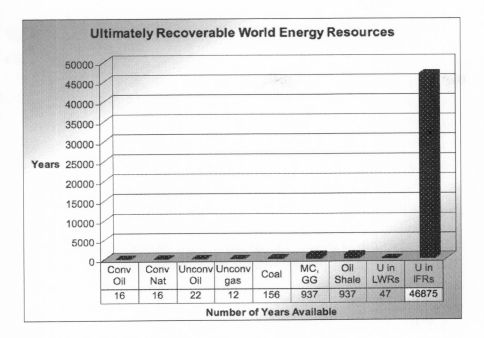

| | | | | | Ultimately Recoverable World Energy Resources | | | | |
Years	Conv Oil	Conv Nat	Unconv Oil	Unconv gas	Coal	MC, GG	Oil Shale	U in LWRs	U in IFRs
	16	16	22	12	156	937	937	47	46875

Number of Years Available

Without getting into the horrendous expense, environmental damage, and technical challenges involved in extracting and using any of the other sources of energy that show more than 16 years, all of which would require carbon sequestration at the very least, it's pretty much a no-brainer, isn't it? How many times have you read or heard some pundit or "energy expert" say that we'll need a mix of technologies, that there's no "silver bullet"? I agree. The bullet isn't made of silver, it's made of depleted uranium. And it would be great if we'd quit shooting it at each other in our endless resource wars and started using it in power reactors instead.

Whatever the expense and extraction difficulties, all the other energy sources shown here are fossil fuels, with all the problems that entails in terms of climate change and other environmental damage. Even if no other energy source were available, uranium alone is sufficient for nearly 50,000 years! Not only that, but the quantities of depleted uranium, nuclear waste, and old weapons we already have available will be sufficient to power the whole world for hundreds of years. We could eliminate not only oil and gas drilling and coal mining, but uranium mining as well.

You owe it to yourself to determine if what I have written about IFR plants is true. (Hint: you can start with the footnotes.) There has been no other concrete solution offered that uses available technology to actually solve the problem of global warming. We desperately need one. This point cannot be stressed too much: All the usual arguments against nuclear power don't apply to IFRs. Quite the contrary. IFRs hold the only solution to eliminating the vast stores of nuclear waste that we've built up so far, usually the first argument brought against nuclear power. According to the International Atomic Energy Agency (IAEA), "Sustainable, environmentally clean long-term use of nuclear power can be achieved only with fast reactors."

Safety and proliferation, the other two most frequently voiced arguments, also lose their dread with an IFR program that is operated in a reasonably sane manner. The simple physics of the materials and the design of the plants make even the worst-case scenario—a terrorist in the control room—unable to cause a meltdown. The plants would be removing dangerous materials—both weapons-grade and that which might be reprocessed into it—from circulation. Even the problems with uranium mining and milling are eliminated, because we'd be able to close down all the uranium mines for hundreds of years. By then it's a pretty sure bet that we'll long since have made fusion power or some as-yet-unimagined energy source commercially viable. If not, I would encourage your distant descendents to take it up with mine. If they decide they need uranium mines to keep the world humming along with IFRs a thousand years hence, they'll have robots by then that can handle it from start to finish.

If you do satisfy yourself as to the advisability of employing IFRs to meet our energy shortfalls and bring our greenhouse gas emissions to an end, then it doesn't matter what you think about renewables, unless you're unrealistic enough to think that renewables alone will meet all our energy needs. Assuming you're anything close to a realist, and that you concur that it's advisable to address the rest of our energy needs with IFRs, then you have to ask yourself just how deeply you want to jump into that end of the pool.

Bear in mind that a wholesale conversion to IFRs would take decades. Any advances in renewable energy technologies and capacities during that time would only be icing on the cake, necessitating fewer IFR plants being built. This IFR plan is not the enemy of renewables. Far from it. This is merely a proposal to replace the portion of the world's energy needs that renewables can't meet with the best technology available, one that coincidentally can solve the thorny problems bequeathed to us by half a century of nuclear stalemate and peaceful yet problematic nuclear power.

So here is my immodest proposal: Let us go light-years beyond the Kyoto accords. For all that they purported to accomplish, they were purely symbolic anyway. The object is not to pave the road to hell with our good intentions. We want to virtually halt humankind's output of pollutants and greenhouse gases, not just slow it down. If we employ only renewables and IFR reactors for our power sources and boron fuel and electricity for our vehicle fleets, we can actually succeed in that endeavor. The sooner we do it, the greater the chances that we'll be able to stop climate change before it becomes irreversible. Perhaps it already has, but we certainly don't want to make it any worse than necessary. At this point we have to do the best we can and hope it's good enough. We might want to put a little funding into the study of earthly terraforming while we're at it. If environmentally benign fusion power reaches commercial viability we can talk about what direction we'll take from there. By that time we should have things under control, climate-wise, at least as much as we possibly can, and the nuclear waste problem will be a footnote in history.

This would seem to be where we should crunch the numbers to see if this all makes sense economically and logistically. But there's one more energy source we haven't looked at yet, a surprising addition to our stable of hydro, solar, wind, and IFRs. Besides supplying us with energy, it takes care of several other pressing problems in the bargain.

CHAPTER SEVEN

Exxon Sanitation, Inc.

Whaddya mean? I'm a legitimate businessman!
I'm a waste management consultant.
— Tony Soprano

I HAVE A CONFESSION to make.

Back in the beginning chapters of this book, I promised to suggest workable solutions for a number of seemingly intractable problems: global warming, nuclear proliferation, nuclear waste disposal, air pollution, and resource wars. Though we're not finished yet, you can surely see the outlines taking shape. Before we go any further I have to admit that I held something back. There are even more problems that we can solve with a few revolutionary technologies and some well-considered political and economic decisions. But how could I list all of them so early in the book without sounding like a utopian dreamer? Hopefully by now you'll know that while yes, I'm a utopian dreamer to some extent (imagining that politicians might actually buck corporate pressures and make sound decisions in humanity's best interest), the dreams have a solid grounding in reality. There is more good news to be considered, even more surprising solutions.

From where we stand at this point in our narrative, some

nagging issues can be seen around the fringes. Boron/electric hybrids sound like they'll put the oil industry out of business except for the still-needed production of lubricants and plastics and the like. Well, there's good news there too, even some good news for the oil companies — despite the fact that oil drilling will indeed cease as surely as will uranium and coal mining. Then there's the nagging problem of man-made greenhouse gas emissions besides the ones issuing from power plants and vehicles. The leading cause of anthropogenic methane emissions in the United States, for instance, is landfills.[152] As we've seen earlier, methane is twenty times more harmful than carbon dioxide, molecule for molecule, in terms of its greenhouse effect (though fortunately it's considerably less persistent in the atmosphere).

Yesterday (as I write this in early 2007) Al Gore addressed the U.S. Congress and declared that by 2050 developed nations should devise a plan to reduce their greenhouse gas (GHG) output by 90%. Just a week earlier Tony Blair was hailed as "bold" for declaring that we should be shooting for 60%, which I guess makes Al Gore really *super*-bold. Yet their boldness seems to have been limited to the goals themselves, since neither of them proposed any solutions that would have even a hope of achieving those targets. Not only that, but Gore's plan stipulated just a 50% reduction worldwide, with the expectation that developing countries wouldn't be able to meet such ambitious goals. Just how the developed nations are supposed to achieve those reductions, however, was never specified, a fairly glaring omission by any standard. As far as this book is concerned, though, not only do we propose to reduce global GHG emissions by upwards of 95%, but we actually have a concrete plan to make it happen (patience, please, you haven't finished the book yet). That leaves politicians like Blair, Gore, and their successors to be, hope-

---

[152] Fact Sheet, "Final Air Regulations for Municipal Solid Waste Landfills," ed. US EPA (Mar 1, 1996).

fully, bold enough to make the necessary decisions. They talk the talk, now let's see if they'll walk the walk. Wishing will not make it so.

If we want to exceed that 95% goal, it would surely help to get rid of landfills, so let's get down to business with that project. Landfills have been a big headache for people all over the world, and out of desperation many municipalities have resorted to burning the trash they generate in giant incinerators. While the proponents of such "waste-to-energy" systems promote them as a solution to the problems of siting garbage dumps, they are far from being a wholly desirable solution. While incinerators undeniably cut down the volume of material to be disposed of, the substances therein and the incinerators' emissions still have to be dealt with.

In earlier years before environmental awareness kicked in, incineration of municipal solid waste (MSW) consisted of simply burning garbage and then burying the ash in a convenient location. That was particularly bad news for the people who lived downwind from the incinerators, though, because the fly ash that was pouring from their smokestacks contained concentrations of heavy metals like mercury, lead, cadmium, arsenic, and copper, as well as some particularly nasty compounds like dioxins and furans.[153] Once this began to be recognized and perceived as a public health problem, a vast assortment of scrubbing devices was invented, and modern MSW incinerators today manage to remove nearly all these substances from the smokestack emissions—when everything is operating perfectly, that is. But we don't live in a perfect world.

For environmentally sound incineration, air pollution control equipment must be serviced regularly by highly

---

[153] Committee on Health Effects of Waste Incineration, "Waste-Incineration and Public Health," ed. National Research Council (National Academy of Sciences, 2000).

specialized personnel. Monitoring equipment is costly and
requires aggressive maintenance and servicing by trained
technicians. In summary, when incineration is done in a
manner that has low adverse health and environmental
impacts it is expensive. When it is done poorly (with low
financial costs) it can be expensive in terms of human
health and environmental impacts.[154]

The difficulty, and the necessity, of maintaining
emissions control systems in essentially perfect order over
a long period of time is daunting even to industrialized
countries. Small mistakes in the operation of such
facilities can easily lead to significant emissions of toxic
substances.[155]

Even if we assume that this sort of scrupulous dedication
to tightly constrained operation and maintenance will be
universally practiced all over the world — oh, never mind.
You know as well as I do that it wouldn't have a snowball's
chance in hell of happening in the USA, nor any other country, much less everywhere in the world. But even if it did, it
only means that the heavy metals, dioxins, and other nasties
end up in the ash, and you have to put the ash somewhere.
This leads to the quandary of figuring out good places to
bury it, often with liners of various kinds, to try to minimize the amount of toxic substances that might leach into
groundwater. Like the expectation of perfection described
above, it can hardly be expected to be universally successful
in every community that has an incinerator. The "waste-to-
energy" concept, by the way, is somewhat of a public relations ploy. While the heat of incineration can be harnessed

---

[154] Larry Rosenberg and Christine Furedy, "International Source Book on
Environmentally Sound Technologies for Municipal Solid Waste Management," (UNEP Environmental Technology Center, 1996).
[155] Ibid.

for a modest amount of energy, there are far less costly methods of producing energy. It's more a matter of making a bad deal a bit less bad.

Fortunately we needn't resign ourselves to choosing between the lesser of two evils, for landfills and incineration are both far inferior to a technology that is just in the early stages of widespread commercial deployment. Any comparison with incinerators would be superficial and quite inaccurate, for the principle involved is not a combustion process. The technology goes by various names — plasma waste conversion, plasma gasification, even plasma reactor. The trick is in the plasma.

Plasma is considered the fourth state of matter, the other three being the more commonly recognized solid, liquid, and gas. When you heat a solid, you get a liquid (in most cases). When you heat a liquid, you get a gas. When you heat a gas, you get plasma. A thermal plasma is an ionized gas that becomes both an effective conductor of electricity and also incredibly hot. We're talking about almost 17,000°C (30,000° for all you Fahrenheit fans). That's a few times hotter than the surface of the sun. Plasma torches have been used for various industrial purposes for years. If you want to cut a twelve-inch-thick piece of steel, you'll want one. They are sometimes referred to as lightning on a stick.

The happy marriage of plasma and garbage promises to make landfills and incinerators mere relics of a bygone age, alongside coal-fired power plants and gasoline engines. But eliminating garbage isn't the only purpose of a plasma converter. Unlike incineration, a plasma converter is actually a recycling device *par excellence.*

Instead of a garbage truck dumping its contents into a landfill, in a more sensible world it will dump it into a giant hopper, from where it will drop through a massive shredder, if necessary, to break its contents down into a reasonable size for a plasma converter to digest. The garbage mélange is then fed into a chamber where the plasma can do its thing. The intense energy transfer that occurs in the plasma is sufficient to rip the molecular bonds asunder, reducing the components of the garbage into their constituent elements. The resulting products exit the plasma chamber as a gas and a very hot molten stream.

The gas that is thus formed is usually referred to as "synthesis gas," or syngas. Its main constituents are hydrogen and carbon monoxide. Syngas is a very useful substance, for it contains all the building blocks of hydrocarbons, from which we derive the myriad petroleum-based products we use every day: fuel, plastics, lubricants, etc. Many people are familiar with synthetic motor oil, which is one of the many products made from syngas today. It is far superior to petroleum-derived motor oil.

When syngas exits the plasma chamber it is understandably very hot (about 1,200°C), and by running it through a cooler a great deal of steam can be generated that can be used to drive a turbine to produce electricity. But of course that leaves us with the syngas itself. It can be burned immediately through a steam or gas turbine to provide substantially more electricity. About 20-25% of this total amount of electricity can be channeled back to run the plasma torches and the plant, while the remaining power can be fed into the grid for sale. Thus a plasma converter for unwanted garbage can become a significant player in the electricity market. If all the U.S.-generated MSW were processed with plasma, by the year 2020 the expected 1 million tons per day of MSW could supply up to 5% of the nation's electrical requirements.[156] This is equivalent to the electricity generated from 25 nuclear power plants. This amount of renewable energy far exceeds the combined energy anticipated from solar, wind and landfill gas projected to the year 2020.[157]

---

[156] Louis Circeo and Kevin Caravati, "Plasma Processing of Msw at Fossil Fuel Power Plants" (paper presented at the HTPP9 Symposium, Orlando, FL, Feb 22, 2007).

[157] Ibid.

If syngas is burned to generate electricity, it will admittedly produce carbon dioxide. However, since the fossil fuel component in municipal solid waste is generally less than 10%, the process is very nearly carbon neutral.[158] It should also be pointed out that by eliminating the inadvertent production of methane that would otherwise result from landfill burial, the situation is improved by several orders of magnitude, since as we've noted before, methane is twenty times more potent than carbon dioxide in its greenhouse effects. Once petroleum is displaced, the percentage of fossil fuel products in MSW will continually decrease until it essentially disappears. At that point burning the syngas to produce electricity will be truly carbon neutral.

There are many other uses for syngas, though, besides just burning it for electricity. Syngas can provide the chemical building blocks for a great variety of products. Methanol can readily be generated from it, at about half the cost of ethanol and in less time. I know this won't be good news to the farmers who've seen their corn prices skyrocket lately due to the heavy subsidization of ethanol plants, but I'm afraid they'll just have to go back to the antiquated notion that farmers grow food for a living. In point of fact the vast majority of taxpayer-funded ethanol subsidies go to giant agribusiness firms, not small farmers. ADM alone rakes in about $1.3 billion dollars/year on ethanol subsidies of 51 cents per gallon. We'd be better off just sending small farmers a check (and telling ADM to pound sand) than to continue subsidizing an energy source with so many disadvantages. It's a greenwashing scam to harvest votes while paying off corporate cronies. As for those who've bought stock in the ethanol plants that are springing up in farming communities throughout America's corn belt, I have one word of advice: sell.

There are other fuels that can be derived from syngas too, with varying degrees of efficiency. Gasoline is the most obvi-

---

[158] Louis Circeo, *Plasma Processing of Msw at Fossil Fuel Power Plants* (Atlanta, GA: Georgia Tech Research Institute, 2007), Poster.

ously usable one. Mobil developed a system to produce gasoline from methanol back in the Seventies. Butanol is another that holds tremendous promise. You may recall the earlier mention of Virgin Fuels, Sir Richard Branson's research project to find a way to power jet aircraft with biofuels. Butanol, a 4-carbon alcohol, is one of the prime prospects. A liquid fuel with an energy density nearly identical to gasoline, butanol can be mixed at an 85% butanol/15% gasoline ratio that will burn in most cars without any modifications to their ignition system. In fact, many older cars can run unmodified on 100% butanol.[159]

Since at least in the near term it may be difficult to engineer boron-powered motorcycles and other small engines, butanol or other alcohols such as methanol and ethanol should be able to fill that gap. It's possible there may need to be slight modifications to a motorbike's ignition system, but clearly the challenges are far from insurmountable, and if necessary they could still run on gasoline derived from waste (and thus carbon-neutral). This would seem to be a minor detail to Americans, but in developing countries around the world the motorbike is a ubiquitous mode of transport. For those of you who haven't had the opportunity to travel in such countries, many of them have a lot of something else that's ubiquitous besides motorcycles: garbage. Butanol on the hoof.

Between boron and butanol, the oil industry is looking more and more like a dinosaur. Plastics, anyone? Sorry, Exxon, syngas from garbage will provide all the chemical components for plastics that petroleum now provides. Synthetic motor oil from syngas is already well on its way to displacing petroleum-derived oil, especially as the public comes to understand its clear superiority (including the fact that it has to be changed much less often). Oh, and speaking of motor oil, we'll be using a lot less of it, if any, when we switch to boronmobiles. As for all the

---

[159] David E. Ramey, *Butyl Fuel, Llc* ([cited 2007]); available from http://www.butanol.com/index.html.

holdouts who'll be driving their old gas guzzlers (now burning butanol, or garbage-derived—and thus carbon neutral—gasoline), when they change their oil they won't have to take the old drain oil to any special place for disposal. Just throw it in the trash and it'll end up in a plasma converter, ready to be made into new oil or any number of other hydrocarbon-based items.

With all the uses to which syngas can be put, let's not forget there's also the molten waste stream emanating from the plasma chamber. This can itself be used in a variety of ways, and like syngas it will contribute to the profitability of the plasma plant. From the molten state it can be spun directly into rock wool, a substance rather like fiberglass that can be used in much the same way. Since rock wool made in this manner would be considerably less expensive than fiberglass, much more insulation can be added to a structure for the same cost as fiberglass, reducing the energy demands of cooling and heating. Compared to the cost of making rock wool the old-fashioned way, spinning it out of the molten slag stream of a plasma converter will cost about one-tenth the price.[160] Rock wool is also lighter than water and highly absorbent, so it can be used to clean up oil spills. And what, pray tell, would one do with the oily mess of rock wool after such an episode? Drop it into a plasma converter, of course. But this particular application will have limited utility, because hopefully the oil industry will be nothing but a memory in the very near future. There won't be any more oil spills to clean up. Ever. No more black feet after walking those idyllic beaches in the tropics, either. Sometimes it's the little things that make it all hit home.

If the molten slag stream is water cooled, nodules of mixed metals can be recovered. These can be sent to metal refineries and effectively "mined" for their component elements. Thus not

---

[160] Jonathan Strickland, *How Plasma Converters Work* (2007 [cited May 3 2007]); available from http://science.howstuffworks.com/plasma-converter2.htm.

only iron, aluminum, and other useful metals can be recovered, but heavy metals from the waste stream that have been such a problem with current methods of waste disposal can also be isolated for reuse.

The slag that's left will be comprised mainly of silicates and other minerals, which can be used for tiles, bricks, roadbeds, etc. But if all the garbage in the world is being run through plasma converters, it will provide so much in the line of these building materials that it begs the question of what to do with the excess. Since the molten stream can be simply allowed to cool into a vitrified (glassy) substance that is nearly inert and highly resistant to leaching, it would seem that simply burying it would be a reasonable course of action. But why not put it to better use?

If any plasma plant found itself with a saturated building materials market and had to look at disposal of the slag, the simple expedient of having some molds handy would be ideal. The molten slag could be poured into molds of various shapes optimally designed for use as artificial reefs. There's an organization[161] that has been manufacturing what they call "reef balls" for some time now, constructing artificial reefs around the world. (Currently they're made of concrete, but slag would work great.) Whereas it may seem logical to think that this sort of thing might make sense in the tropics where coral reefs are most commonly found, one need only look at artificial reef projects off New Jersey or even farther north to see that the range of possibilities is nearly endless.

While the sea's seemingly limitless bounty might lead people to believe that it's teeming with life, the truth is that the vast majority of the sea bottom is relatively featureless and barren. All along the continental shelves there stretch seemingly limitless expanses of relatively smooth terrain with a minimal

---

[161] *Reef Ball Foundation* ([cited 2007]); available from http://www.reefball.org/.

amount of animal and plant life. But drop a pile of nearly any solid material onto the bottom and watch what happens. As soon as there's something to anchor to, planktonic organisms like barnacles, corals, sponges, sea squirts and others will come floating by and latch on. Crustaceans will make their homes in the nooks and crannies. Fish will arrive and take up residence. Pretty soon you've created an entire community, a little neighborhood ecosystem where virtually nothing lived before.

Human communities that have created artificial reefs offshore have seen them generate not only fish but dollars. Sport fishermen who had no reason to visit before now suddenly find good fishing. Even commercial fishing is enhanced, especially where extensive reef building has resulted in ever more diverse fish populations. In a time when pollution and destructive fishing and mining practices have damaged or utterly destroyed natural reefs in many parts of the world, this possibility of dramatically increasing the biological carrying capacity of continental shelves is a golden opportunity. Permits to dispose of reef-ready slag could be issued to plasma operators by local boards using the advice of marine biologists hired to advise on the optimum locations and volumes of artificial reef materials. Who would ever have imagined that our garbage could be put to such good use?

But like nearly any new idea, there are already groups of individuals lining up in opposition to plasma converters. It's a classic case of Voltaire's maxim, "The perfect is the enemy of the good." The anti-incineration forces mistakenly (or disingenuously) regard plasma converters as simply a sneaky kind of incinerator.[162] Facts take a back seat when delusion is driving. One such group promotes a waste-free society, where everything we use would be recycled — the old way. I suppose we're all expected to have a dozen different waste bins in our homes

---

[162] "Incinerators in Disguise," (Global Alliance for Incinerator Alternatives, April 2006).

into which we'll dutifully sort all our waste — oh, and don't forget the compost heap and the composting toilet. Never mind the fact that plasma converters represent the zenith of recycling without any effort on the part of those creating the garbage. But hey, there's a cause for everyone, isn't there? There's even a Russian group that's agitating to *increase* global warming gases, presuming it'll make Russia far warmer and more habitable.

Getting back to that composting toilet idea, though, brings up yet one more use of plasma converters. The sludge resulting from waste treatment plants needn't be simply buried anymore. That too can be run through plasma converters. There is already a small commercial plasma plant in Japan processing 17 tons of MSW and 4 tons of sewage sludge per day. Pretty much anything can go in there, including dirt. That's no small advantage, since there are many unbelievably trashy places where a front-loader could just drive in and start scooping up trash. In many developing countries there is virtually no garbage collection infrastructure, and residents have no compunction about littering. It seems inadequate, though, to use the relatively innocuous term "littering" to refer to the practice of just tossing every bit of your garbage into the street. Words fail me. Sometimes it just has to be seen to be believed. Some archeologists have suggested that ancient cities are found underground because they've slowly been buried in garbage.

Sometimes natural or man-made disasters resulted in cities being buried more suddenly, though. Looking at the aftermath of Hurricane Katrina in New Orleans, it's not difficult to imagine similar calamities resulting in the total abandonment of ancient cities. The post-Katrina cleanup will likely take years, what with the mix of chemicals, plant and animal debris, and destroyed buildings, all stewing in the heat and humidity. It's a pity we've not quite reached the plasma converter age yet.

One can easily imagine some enterprising business person taking plasma converters to the road in the not-too-distant future. Such a system built aboard one or two large trucks could follow harvests and other intermittent waste-producing events around the country, but at disaster areas is where they would be truly welcome. If New Orleans had had a couple plasma converters already operating nearby to recycle the city's normal MSW load, cleanup could have started immediately, and mobile rigs could have showed up to accelerate the process. It matters not what sort of mixed-up mess of hazardous or more benign material has to be cleaned up. Plasma converters can handle it all, and even make money in the process. Apparently this missed opportunity was not lost on the city fathers of New Orleans, for they're the second city in the USA to announce plans to build a plasma converter.

Looking ahead to its many applications, the profit potential of plasma conversion is tremendous. Private companies could build facilities in developing countries and it would naturally be in their financial best interest to develop the garbage collection infrastructure to support their business. This is a perfect niche for the oil companies. The capital investment is fairly substantial. A plasma plant capable of processing 2000 tons per day — about the amount that a million people produce in the USA (likely less elsewhere, and much less in most places) — would cost about $250 million. The payback time on that investment would vary depending on what the syngas and slag would be used for, but current estimates are about twenty years.[163] This can change considerably, though, since there are so many different uses for the syngas and slag. Also, that payback time is premised on the cost of building the first large plasma converter in the world. The first one always costs the most, of course. The price will surely drop substantially with future construction.

---

[163] Lynne Sladky, "Florida County Plans to Vaporize Landfill Trash," *USA Today* 9/9/2006.

You may be wondering at these rosy predictions if the first big plant is still in the planning stages. But this technology is already in use and has been for some time. Most of the plants have been modest in size and often built for specific uses. Hitachi has been a leader in the field, having built a plant in 2002 that processes both MSW and automobiles shredder residue (ASR). That plant now processes 300 tons of MSW and ASR per day with two plasma units. Other more specialized converters have been used for everything from nerve gas to munitions, and now the U.S. Navy has begun to put them on ships to solve the problems of waste disposal at sea.

The project everyone's watching is a new plasma converter planned for St. Lucie County, Florida, due to begin operation in 2009. This will build upon the experience of Hitachi's plants, scaling up the 150-ton/day gasifier units to 500 tons/day, with up to six plasma torches in each. The plan calls for several such reactor modules capable of handling 3,000-3,500 tons of MSW per day. Since St. Lucie County has a population of about 250,000, they'll easily be able to process not only their own MSW but also that of several surrounding communities. But they'd also like to get rid of their landfill that's been such a problem to them, so they're planning to use some of their extra capacity to gobble it up, bit by bit, until it's gone in about 18 years. The 120 MW electrical output to the grid should be sufficient to power every household in the county. The old adage "One man's trash is another man's treasure" will soon be demonstrated in spades at St. Lucie County.

Another plasma plant is due to come online in Pennsylvania in April of 2009.[164] This plant will process MSW and agricultural waste, and is being designed to produce ethanol at a cost of about a dollar a gallon. Plans call for building 20-25 plants per year to produce "a couple billion gallons" of ethanol

---

[164] Thomas Olson, "Gamble on Plasma Turns into Jackpot," *Pittsburgh Tribune-Review* Apr 26, 2008.

annually, more easily and cheaply than that produced from corn today, without resorting to food crops or special plantings for the purpose. Obviously this will be highly competitive with the cost of gasoline even without any subsidization. One wonders, then, just how long it will take the federal government to stop subsidizing Archer Daniels Midland. Any bets?

At the moment, oil companies are sitting on a stash of about $2.35 trillion and growing.[165] Let's assume for a minute that they keep the 0.35 trillion aside for buying garbage trucks and dumpsters. Two trillion bucks would pay for about 8,000 plasma plants at the prices quoted above, but of course those prices will plummet as soon as they start being mass-produced. It would hardly be a stretch of the imagination to guesstimate that their cash stash could build 15,000 plants, enough to handle the MSW of some fifteen billion people who crank out trash as prolifically as Americans. Since there are just a bit more than six billion people on earth right now, and many of them generate precious little trash (or anything else, for that matter), it would seem that the companies that have heretofore been providing us with our oil are perfectly positioned to become the planet's garbage kings while using a mere fraction of their savings to do it.

This would seem almost too ideal, for who has more experience in the multitude of ways to manipulate hydrocarbons than the oil companies? They could possibly even convert some of their existing refinery equipment in the service of syngas manipulation. Since all the elements would be originating from non-fossil fuel sources, there would be no cause for concern about GHGs. Emissions would be carbon neutral any way you slice it. So have at it, Exxon. Go nuts, BP. Enjoy your new business. And while you're at it, put a little money aside for anti-littering social engineering.

---

[165] Greg Palast, *It's Still the Oil* (3/18/2007 [cited); available from http://tinyurl.com/2vuesr.

Baby boomers know what I mean. Back in the day it was pretty standard procedure to toss stuff out the window as you drove along. Of course in those days most food wrappers and cups and such were made of paper. But it took a while for a consciousness of littering to take hold, until today those who litter are pretty much considered boors in most developed nations. Once garbage infrastructures are in place, it'll be to the advantage of the sanitation companies to inculcate that consciousness worldwide, since it's a lot easier and more efficient to collect trash from bins than to pick it up off the roadside piece by piece. A heck of a lot more pleasing to the eye, too.

As for the tens of thousands of landfills that are already closed or full to bursting, emitting methane and often contaminating groundwater, there is an alternative to digging out all that old garbage. Using a process called In-Situ Plasma Vitrification (ISPV), boreholes can be drilled into old landfills and cylindrical plasma torches can be inserted deep underground. As the plasma gasifies and melts the material below, syngas can be drawn off at the surface. All the solids will remain underground, cooling into an inert glassy slag, effectively entombing the troublesome heavy metals and other substances that have been a continuing pollution hazard for groundwater supplies. If the boreholes are spaced closely enough, the molten pools will coalesce into a solid layer, and as the plasma torches are slowly raised to the surface the molten layer will be transformed into a glassy underground monolith. While the ground can be expected to subside substantially during the process, the end result will be a completely stable surface with none of the pollution concerns of the past. The syngas that has been drawn off can of course be used however the "miners" prefer. When the ISPV process becomes fully developed, it may become very cost-effective to mine existing landfills for energy production.

This process of drilling boreholes holds more than pecuniary promise, however. Remembering that plasma conversion breaks down compounds into their constituent elements, consider the many severely polluted sites that desperately need to be cleaned up to prevent groundwater contamination and illness to nearby residents. In the USA these have been designated as Superfund sites, named for the pile of money that Congress has intended to allocate for solving the most serious localized pollution problems.

Most of these sites are polluted with toxic chemical compounds. The plasma torches, burrowed deep into the problem areas, would break down most of these compounds into their harmless constituents. In cases where some of the elements themselves are problematic, such as heavy metals, they would end up tightly bound in the vitrified slag, impervious to leaching for thousands of years. Whereas most Superfund sites today are cleaned up by hauling out untold truckloads of contaminated soil — at tremendous cost — plasma conversion would accomplish the task much more effectively on site. After all, when contaminated soil is trucked away it's usually a matter of trying to make a bad situation a bit less bad by finding a less sensitive place to dump it. With plasma conversion, though, there is no waste product to be disposed of at all. The cost of cleaning up these pollution hotspots can be reduced by an astounding amount, and be accomplished quickly to boot. Up to now the cleanup has been hampered by both budget constraints and devilish technical challenges. Neither will be an issue once plasma technology is employed.

Even though plasma converters cannot transmute radioactive contamination into non-radioactive elements, the system was nevertheless called upon in an attempt to clean up the ground contamination at the Savannah River nuclear power plant in Georgia. This effort, under the direction of the U.S. Department of Energy, succeeded in entombing all the radioactive elements in a vitrified underground mass, which prevented its spread through the water table and essentially nullified the problem. The DOE, having been rather desperate to solve that prickly dilemma, declared ISPV to be ready for commercialization.[166] The technology clearly demonstrated its ability to convert the most problematic hazardous, toxic, and even radioactive wastes and contaminated soils into stable, vitrified forms.

The effectiveness and economy of harnessing ISPV for Superfund cleanup can transform that program from a lumbering, costly and procrastinating beast into an efficient and profitable enterprise. Despite its misleading moniker, the Superfund has been consistently starved for cash and the cleanup of over 1,200 sites around the USA has proceeded in pathetic fits and starts. At last we have a means of accomplishing this formerly daunting task by harnessing the power of plasma.

If oil companies decided to go whole hog into the garbage business, and even went so far as mining old landfills, their total investment would still leave at least hundreds of billions of dollars, probably over a trillion, just sitting in their coffers waiting to be used. Fortunately for them, municipal solid waste represents just a small portion of the total waste stream. Far more material is available in the form of industrial wastes, agricultural waste, and construction debris. Many of these materials have the unfortunate characteristic of being hazardous in one degree or another, and the industries that produce them as a byproduct of their operations pay substantial sums to have them disposed of by companies that specialize in such operations. Undoubtedly you've heard of many cases where such hazardous waste "specialists" have surreptitiously dumped their cargo at sea, in rivers, in landfills, or shipped them to unfortunate countries in the developing world where they've sickened

---

[166] P.G. Zionkowski R.F. Blundy, "Final Report, "Demonstration of Plasma in Situ Vitrification at the 904-65g K-Reactor Seepage Basin."", ed. DOE (Aiken, SC: Westinghouse Savannah River Company, Dec 1997).

or killed the hapless residents.[167]

Once plasma converters are widely deployed, hazardous waste disposal prices will drop substantially, since the process of dealing with them will be greatly simplified. Like the other inevitable economic casualties along the way to our new energy paradigm, the illustrious employees of Slippery Tony's Midnight Hazardous Waste Disposal will have to find honest jobs. I hear Exxon Sanitation is accepting applications.

As for agricultural waste, it's usually preferable to recycle it directly into the soil even though it could be converted to usable materials in a plasma converter. But there will be no shortage of raw material beyond garbage to make plasma conversion a burgeoning growth industry in the very near future. Virtually any byproducts of industrial or agricultural processes that are now discarded will be candidates for transmutation into beneficial products. Syngas and metals will be the prime values. One can't help but think that the mainly silicate slag will ultimately be so abundant as to make building and paving materials ridiculously cheap. But with judicious reef building programs, the slag that ultimately finds its home along the margins of our continents may end up indirectly translating into one of the most valuable products of the plasma systems.

It's even possible that plasma conversion could turn out to be the most direct and economical method of recycling boron oxide for our automobile fleet. Given plasma's ability to sever molecular bonds, it seems reasonable to suggest that feeding boron oxide into a plasma converter would result in the oxygen being liberated as a gas while the boron reverts to its pure elemental form. If so, then the oxygen could be drawn off and combined with hydrogen from an adjoining plasma converter that's busy with the task of converting garbage, agricultural waste, or any other organic materials. The hydrogen-oxygen

---

[167] Zada Lipman, "A Dirty Dilemma," *Harvard International Review* 23, no. 4 (Winter 2002).

combination produces a prodigious amount of heat (it's used to fuel the space shuttle) that could be used to run a steam turbine and generate electricity to power both plasma burners. Unlike relying on other sources of energy to create steam via a heat exchanger, the combination of oxygen and hydrogen creates not only plenty of heat but water—conveniently in the form of steam. Such a system would preclude the use of electricity from the grid for the boron oxide recycling process, instead deriving its energy from the incoming streams of oxygen (on the boron oxide side) and hydrogen (on the garbage side).

In such a scenario, drawing the hydrogen off from the syngas on the garbage side would leave mainly carbon monoxide. If that was simply liberated into the atmosphere it would be carbon neutral (being made from organic materials rather than fossil fuels). But rather than releasing it, why not sequester it underground? After all, aren't we all being led to believe that carbon sequestration is the answer to our continued use of coal? If it's so feasible, then here's where it could be employed to chip away at our atmosphere's GHG problem in a big way, for the organic materials feeding the plasma converter would have derived their carbon from the atmosphere. The process of carbon sequestration, as proposed by the coal industry, involves deep drilling into stable formations. Who knows how to do that from a long history of having already done so? Sure, the oil companies, now in their new incarnation as garbagemeisters. They've even got the drilling rigs to do it, which will be sitting around rusting in a modern version of a Halliburton nightmare. They might as well put them to some positive use.

While wholesale conversion to all-electric households can proceed quite smoothly in industrialized countries, developing nations lacking extensive electrical transmission grids will be far more dependent on liquid fuels for cooking and other energy needs. Boron will be able to fill the bill to some extent, but the generation of methanol, butanol, and other fuels from plasma

converters will be invaluable in converting the energy infrastructure of developing nations to environmentally sound systems, even before their electricity grids are built. The ease with
which methanol can be produced from syngas is especially welcome, since it will provide an inexpensive and easily transported
fuel for cooking stoves. It would be well worth it for the developing nations to subsidize methanol for these purposes, perhaps
with a modest tax on either boron or electricity. This is not a
purely altruistic notion. A large part of the Asian Brown Cloud
is made up of particulates from the dung or wood cooking fires
of millions of poor people, who also suffer horribly from the indoor pollution that only belatedly makes its way outdoors. The
sheer number of people using such cooking methods creates a
pollution hazard that respects no international boundaries.

Aiding the most impoverished among our planetary brethren isn't the only guilt relief that the many benefits of plasma
technology will provide, however. Parents will be able to diaper
their babies with disposable diapers, knowing that the diapers
(plus their bio-cargo) will all be converted into usable materials.[168] My son — long since out of diapers — even came up with
the concept of the Guilt-Free Car, made almost entirely out of
garbage. Rock wool spun from molten slag will take the place
of fiberglass in a car body that obtains most of the elements
of its accompanying polymer resin from syngas. Likewise the
tires and plastics used throughout, even the upholstery fabrics,
would be derived from syngas. Metals can be recovered from the
nodules produced by the plasma factories. Even the upholstery
padding can be made of rock wool. The Guilt-Free Car will run
on boron, of course, which may even end up being recycled using the energy from MSW or other waste products. Now there's
a car any environmentalist would be proud to drive.

---

[168] American babies alone use about 18 billion disposable diapers a year,
and their use is increasing rapidly around the world.

Plasma converters represent the ultimate in recycling, making virtually 100% of the waste a household normally produces into usable and even valuable end products. There would be no need to have two garbage pickups every week, one for trash and one for recyclables that people have perhaps been conscientious enough to separate. Everything could go in the trash. One might wonder about glass, though, because whereas the mainly silicate slag is itself a glassy substance, it couldn't be used to make glass containers because it would be mixed with other minerals. On the other hand, silica (from which glass is made) is the most abundant mineral on earth, so even if people don't sort their bottles it wouldn't really be that big a deal. It's not like we're going to run out of sand. We can make all the bottles, jars, and windows we want, and we'll still have plenty left over for important things like computer chips and breast implants.

As for the ex-oil companies, now kings of all the garbage they survey, there would not only be a profitable business, but they could quit worrying about peak oil. There will never be a peak garbage point, at least not until the human population of the earth starts to shrink. Not only is that population, alas, still growing by leaps and bounds, but as prosperity spreads people end up producing more garbage. Yet since everything can be so thoroughly recycled, there's no need to be overly concerned about it.

Those who feel virtuous about sorting their garbage, driving a hybrid, and wearing a sweater so they can keep the thermostat down will just have to find other reasons to feel virtuous. With free depleted uranium providing unlimited cheap electricity to heat (or cool) everyone's home, if you don't mind paying a little more on your electric bill every month you can feel free to doff the sweater. You can toss whatever you want in your single garbage can, then run outside and jump in your boron-powered SUV and cruise away. In such a future there will be one thing in short supply, though: guilt.

Does this all sound too good to be true? Sure, the plasma converter and all its offshoots clearly constitute a viable enterprise in and of themselves, but if we're talking about building thousands of IFRs the costs must be astronomical. If we want to run an all-electric world powered by IFRs with boron-fueled vehicles and plasma converters working away, just how many reactors will we need and how much is it all going to cost?

WARNING!

The following chapter may induce big number vertigo.
Enter at your own risk.

CHAPTER EIGHT

Check, Please!

A billion here, a billion there —
pretty soon you're talkin' real money.
— Everett Dirksen[169]

MANY PEOPLE WHO used to reject nuclear power out of hand have come to the conclusion that the threat of global warming is so grave that they're now willing to embrace the nuclear solution, even thermal nuclear plant technology with its problematic waste and other issues. Yet converting humankind's entire energy production over to nuclear power (with whatever solar, wind, hydropower and other clean technologies can add to the mix) is a goal that has rarely been broached. But does a partial solution make sense? Let's look at this a little more starkly before we start crunching numbers.

As to solar energy, it is a wonderfully clean and ubiquitous source of endless power. Nature's fusion plant is cooking away right there in the center of our solar system (I'm assuming even the global warming skeptics agree with Copernicus here). The main problem with collecting it for generating electricity,

---

[169] Someone once asked Dirksen about this famous quote. He replied, "Oh, I never said that. A newspaper fella misquoted me once, and I thought it sounded so good that I never bothered to deny it."

however, is that it is so diffuse. Only a certain amount of solar radiation falls on a square meter of earth's surface under even the best of conditions, and none at all, of course, at night. So the amount of sunlight that we can convert to electricity is limited by four things: how much sunlight is hitting a given surface, how efficiently our collector can convert that to electricity, how much acreage we're willing to devote to collecting it, and how much it would cost.

The amount of solar radiation varies slightly depending on solar activity cycles, but for our purposes we can safely assume that it's constant. Many areas of the earth are poorly suited to solar electrical generation because of their latitudes or frequent cloud cover. Then there's that problem of efficiency. Our necessarily brief look at solar in Chapter Two, even as we considered state-of-the-art facilities in some of the best locations in the United States (speaking strictly from a solar power point of view) amply illustrates the point we've reached with efficiency. There is room for improvement, of course, and hopefully a lot of improvements will be discovered. But the climate change clock is ticking. Even if efficiency can be dramatically improved, the very dilute and variable nature of sunlight on the earth would still require such massive deployments and economic commitments that the very idea of converting the world's entire energy infrastructure to solar is beyond reason. Long before such a thing happens we could even see fusion power going commercial. Meanwhile, acknowledging that improvements can be made and that some of the planet's energy load can be shouldered by solar, it would certainly seem wise to continue its development. But don't expect it to pull our fat out of the fire, not by a long shot.

Similar issues are there with wind, as we know, and I want to reiterate the fact that I do not wish to diss the solar and wind proponents, nor to exclude them from the energy picture. Far from it, in fact. I dare say most of us would be tickled pink if these options were able to be scaled up to provide all or even a major portion of the world's energy needs. We know it's not about to happen in the foreseeable future. Hydro is probably going to be relatively static in much of the developed world due to resistance to dam building. Indeed, the failure of salmon runs in the western United States has people calling for dam removals. As for biofuels, the other major piece of the renewables puzzle, in the near term the fuels we could derive from plasma converters and garbage seem to offer the most logical and economical option. Plasma technology also avoids the environmental damage being done by a pell-mell rush to biodiesel and ethanol that's already wreaking havoc with sensitive ecosystems and global food prices. Indeed, plasma converters promise considerable environmental benefits.

The premise of the energy revolution I'm proposing is that IFRs represent the best option for providing the vast majority of the primary energy that humanity needs until the dawn of the fusion age, however far off that may prove to be. If that is the case, and if all non-renewable fossil fuel power sources are inferior to it (and dangerous to our planet's health), then logic would seem to dictate that our best course of action would be to convert all our non-renewable energy production to IFRs. A thousand or so nuclear plants (as the aforementioned MIT study and many others envision) is simply another half-measure, and half-measures just can't solve our very serious problems. As a matter of fact, the situation will just continue to get worse with such modest steps. If you don't believe that IFRs are our best non-fossil fuel power option after educating yourself enough to know (and you'll learn much more about them by the time you finish this book), then by all means please suggest a better one. If you do believe they are the best option — even if you don't like that conclusion on an emotional level — then don't let big numbers keep you from what reason would suggest: an IFR building plan of Manhattan Project-like urgency to shut down the inferior power systems all over the world that are contributing to our climate change dilemma.

The most pressing issue is to shut down all coal-fired power plants and usher the coal era to an end. The oil industry can be phased out even more quickly with a commitment to boron hybrids, and from both a geopolitical and an environmental standpoint that in itself would be a boon to humankind. But with coal contributing even more GHGs than vehicle emissions on top of its staggering socioeconomic and environmental costs, shutting down coal plants should be high on the global agenda. Natural gas will be the last of the fossil fuel industries to be phased out, and as soon as thermal nuclear plants reach the end of their lifetimes uranium mining and processing will follow. When all nuclear power plants are IFRs, there will be no need for uranium mines for centuries, and the thorny problem of long-lived nuclear waste will have been solved once and for all.

So what kind of money and timelines are we talking about here? As to the latter, the idea of building hundreds of nuclear plants a year is something I haven't seen even remotely suggested by anyone, though there are really no compelling reasons, given the political will, that it couldn't be done. France has been good enough to give us a perfect demonstration.

Once the oil shocks of the early Seventies jolted the world into a new perspective, France more than any other nation took decisive action. Having precious few natural energy sources of its own, the nation embarked on an ambitious plan to convert their energy infrastructure to nuclear power, supplemented by what hydroelectric power they'd already developed. Within the space of about 25 years they succeeded, and today France's fourth largest export is electricity. About eighty percent of their electricity is provided by nuclear power, with nearly all the rest comprised of hydroelectric and other renewable sources. It is truly ironic — and more than a little ridiculous — that France is singled out for being so far behind on meeting the EU's renewable energy target, a system that was put in place to encourage its member nations to reduce their GHG emissions. The fact that nearly all of France's GHG emissions come from the transportation sector and that they produce far lower emissions from their electrical generation systems than any other EU nation just isn't recognized under the renewable energy goal system. So if you happen to see France being castigated as a global warming slacker, take it with a large grain of salt. They are, in fact, helping their neighbors reduce their GHG emissions by selling them electricity from France's nuclear and renewable energy power plants, all the while enjoying the clearest skies in the industrialized world.

France's nuclear power buildup proceeded at the rate of up to six new power plants a year. As in most other countries, they tend to build them in clusters of three or four, with a total capacity per cluster of 3-4 gigawatts electrical (GWe). Currently the government-owned electrical utility, Electricité de France (EdF), operates 59 nuclear plants with a total capacity of over 63 GWe, exporting over 10% of their electricity every year (France is the world's largest net electricity exporter). Their electricity cost is among the lowest in Europe at about 3 eurocents (or €ents, if you'll allow me to coin a new symbol of sorts, since I know of no euro-native symbol akin to the U.S. ¢) per kilowatt-hour.[170]

Herein lies an exceedingly sticky bone of contention, and one we must deal with before we figure out the price tag for the newclear/boron/plasma revolution. On the one hand we have the anties' oft-repeated chestnut about how the utilities used to promise back in the day that "nuclear power will be too cheap to meter." This is usually accompanied by horror stories about how much of our tax dollars have been poured into subsidizing the nuclear power industry and how without being propped up by

---

[170] Uranium Information Center (UIC), "Nuclear Power in France," (Melbourne: Australian Uranium Association, April 2008).

the government nuclear would never be financially viable. The most authoritative sources of information on international energy statistics, like the Organization for Economic Cooperation and Development (OECD), the International Energy Agency (IEA) and various UN bodies and governments, are often implied co-conspirators in some sort of international plot to distort the facts against solar or wind power. If this is indeed the case, then we're stuck with the conspiracy, because their facts — like them or not — will be the basis for political decisions on our energy future. Therefore I will use their statistics to try to sort through the hype and misinformation, while freely admitting that I don't believe for a minute in any such conspiracy.

On the other hand, we do have several decades of actual experience in a large number of countries with a variety of nuclear power programs to actually give us a very good base of raw data to work with in trying to ascertain the truth by means other than polemics. While I would not dispute the fact that the nuclear power industry has received both overt and hidden subsidization (more on that in the following chapter), a dispassionate look at the facts worldwide should provide us with a basis for the cost calculations we'll need. Let it be said up front that these calculations posit that the percentage of electricity provided by renewables will be assumed to be no greater than today, despite the fact that many governments have the intention of drastically increasing those percentages. If that transpires, so much the better, it would mean that fewer IFRs would be needed. From an IFR cost standpoint, then, we'll be using a worst-case scenario, just to be conservative.

France is preparing to begin replacement of their aging reactor fleet with a new design known as the European Pressurized Reactor (EPR), the prototype of which is being built in Finland. This is the first so-called Third Generation reactor in the world, incorporating safety and efficiency improvements intended to set a new standard. The U.S. version of the EPR[171] being considered for future deployment is described as "simple, using 47 percent fewer valves, 16 percent fewer pumps, and 50 percent fewer tanks as compared to a typical plant (four-loop reactor system) of comparable power output."[172] Even at that, such a power plant employs four independent emergency cooling systems, each capable of cooling down the reactor after shutdown.

Compare such a pressurized water reactor to an IFR of the Argonne type. With an IFR you eliminate the four emergency cooling systems right off the bat, because the physics of the IFR's materials and the reactor design itself ensure the plant against coolant emergencies. Since the IFR operates at normal atmospheric pressure, the number of valves and pumps and tanks is reduced to a mere handful. The only pressurized area is the steam portion of the system in the turbine room, which is isolated from the reactor in a separate structure and contains no radioactive elements.

The passive safety concept pioneered in the IFR has been adapted to a new generation of LWRs as exemplified by the Westinghouse AP-1000 reactor and GE's Economic Simplified Boiling Water Reactor (ESBWR). The former has already been certified for production with orders in hand and two units currently under construction in China, while GE's ESBWR is in the final stages of certification. As the first EPR is being built in Finland, its delays and cost overruns (not unusual in prototypes) are emboldening those in France who feel that the EPR has already been superseded by superior designs and should perhaps be abandoned in favor of passive systems like the new LWRs. These promise both safety and economic advantages over

---

[171] Americans have co-opted the EPR acronym to signify Evolutionary Pressurized Reactor. Those insidious Darwinists are everywhere!

[172] Areva, "EPR Fast Facts," in *Unistar Nuclear Energy* (2007).

designs like the EPR because of the aforementioned simpli-
fication that passive systems allow. A glance at comparative
schematics of these designs illustrates the point:

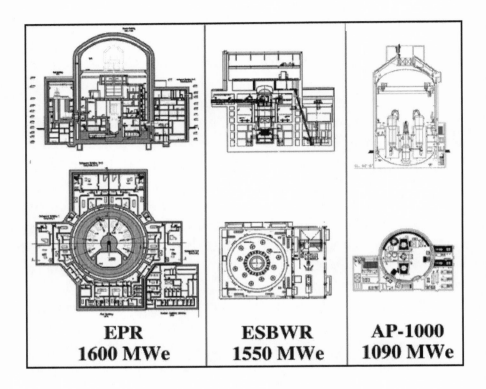

| **EPR**
1600 MWe | **ESBWR**
1550 MWe | **AP-1000**
1090 MWe |

The OECD recently published a study comparing costs of
generating electricity from various sources. Unfortunately
this study didn't include any passive safety designs. Nevertheless, the EPR was clearly the cheapest producer of electricity, even compared to the simple and environmentally
catastrophic pulverized coal power plants. The study took
all relevant factors into account including investment, operation, maintenance, fuel costs, backup systems, interest rates,
etc. The cost for electricity from an EPR swept the field
at €23.80/MWh. Renewables such as run-of-the-river hydro,

onshore or offshore wind, biogas, and various solar systems all came out over twice as expensive, with rooftop PV panels about 15 times more costly.[173] But IFRs (not included in the study), when fueled solely with depleted uranium, eliminate fuel costs as well, aside from the minor unavoidable costs of the metal cladding and other elements of the fuel assemblies. The actual fuel, the depleted uranium, is free, eliminating all concerns of fuel price volatility.[174] If IFRs could be built for about the same price per gigawatt as EPRs (and there's every reason to believe they can be, especially when we build thousands of them), the cost for electricity should certainly not exceed — and might well reduce — the already rock bottom price of EPR electricity.

Yet there is still talk about how nuclear power will be just too expensive and thus commercially impractical, even as France already has plans to completely replace their current reactor fleet with EPRs. Do these naysayers think the French are imbeciles? France already has the lowest electrical rates in Europe and electricity is their fourth biggest export. The contention that they're losing money hand over fist because of the high cost of nuclear plants is absurd on the face of it, yet we continue to hear that fallacy repeated ad nauseum as a rationale for abandoning nuclear power.

It should be duly noted that much of the cost of building nuclear power plants in the United States has been due to two factors that should be substantially minimized or eliminated in the future. Those two factors are a lack of standardization in plant design and the considerable cost of delays and changes

---

[173] International Energy Agency (IEA), "Projected Costs of Generating Electricity – 2005 Update," (Paris: OECD, IEA, NEA, 2005).

[174] Early on, when using up the spent fuel from thermal reactors, fuel price volatility will likewise be eliminated but there will be the added cost of reprocessing the spent thermal fuel to make it into IFR-compatible fuel assemblies.

due mostly to legal action and stalling tactics by anties.[175] I was darkly amused at an article I read recently where an environist was condemning the industry for hiding the exorbitant costs of nuclear power when in almost the same breath he boasted about how his compatriots had forced delays and substantial added costs to the project by dint of legal maneuvers.

The issue here is legal and legislative more than logistical. Given the political will — which would include the intestinal fortitude to stand up to the anti-anything-with-the-word-nu-clear-in-it crowd — most of the legalistic impediments could be surmounted with sound legislation. Thus when ground is broken to build a new nuclear power plant its timetable and budget could have a reasonable expectation of being met. The other costly factor, that of building one-off designs, could and indeed must be replaced by the simple expedient of deciding on the best design and using the same plan for all the reactors. France was quite successful in doing this, though their plants did evolve somewhat as technologies improved. In countries that resort to the cookie cutter approach, costs are substantially lower than they have historically been in the USA. Obviously when we're talking about building thousands of reactors, economies of scale kick in. Standardization is clearly the logical course once a superior design is chosen.

So let's see if we can cut through the hyperbole and get a handle on what power plants cost. This exercise has been complicated by the recent free-fall of the value of the dollar and the skyrocketing cost of oil in 2007 and 2008. The calculations used herein are based on estimates circa 2005–2007, when the Euro and the dollar were roughly equivalent. While soaring fuel costs have driven up prices across the board, our comparisons here are still pertinent because the cost of materials to

---

[175] An excellent elucidation of these issues can be found in Professor Bernard L. Cohen's *The Nuclear Energy Option,* Chapter 9, accessible on the web at http://www.phyast.pitt.edu/~blc/book/index.html

fabricate power systems, be they solar panels, windmills, dams, or nuclear plants, are all rising to a similar degree. (You may recall the earlier story of T. Boone Pickens' wind farm and his ever-escalating cost projections.)

When considering nuclear plants it's important to take into account not only the considerable construction costs but also the cost of decommissioning and nuclear waste disposal. Usually a small surcharge is added to the price of electricity from nuclear plants to contribute to those costs, though I don't believe for a minute that the utilities are building in enough for the sort of waste disposal they're contemplating. Why should they, when it'll be stuck in Yucca Mountain (they believe) and the government will have to deal with it? Anybody who believes Con Edison or PG&E will be around in 10,000 years to pick up the continuing tab is clearly delusional. Once we switch to IFRs, though, the problem will be solved, so what is now a hedged bet will turn out to have been more than ample. In fact, the considerable fund that's already been set aside for waste disposal can be utilized now to reprocess today's spent fuel inventories into IFR fuel assemblies.

Remember when we talked about capacity ratios in Chapter Two in regard to wind power (about 21%)? Nuclear plants these days are running at about 90% or more in the USA. France operates at a considerably lower capacity ratio, about 77%,[176] simply because they have so much added capacity built into the system. As soon as electric cars begin to take advantage of the improvements in technology we're starting to see now, France will be in the best position of any nation to avail themselves of the new technology, since they've got the extra juice to charge them. Yes, nuclear plants cost a lot to build, but are quite cheap to fuel and operate, and they produce an awful lot of electricity. IFRs, with their greatly simplified design and modular com-

---

[176] (UIC), "Nuclear Power in France."

| Fuel Mix | $/MWh | | $/kW(e) |
	5% discount rate	10% discount rate	Construction Costs
Coal	25-50	35-60	1,000-1,500
Gas	37-60	40-63	400-800
Nuclear	21-31	30-50	1,000-2,000
Wind	35-95	45-140	1,000-2,000
Hydro	40-80	65-100	NA
Solar	150	200	NA

Figure 1: Source: International Energy Agency (IEA), "Projected Costs of Generating Electricity", 2005[177]

ponents, should be even cheaper to build, and the free fuel just adds to the economic advantages.

The numbers in Figure 1 represent the lifecycle costs for power plants all over the world. These are total costs versus output for the lifetime of the plant, with all costs included: construction, operations and maintenance, fuel, and decommissioning.

Nuclear clearly is the economical choice here, though as mentioned earlier renewable energy proponents will sometimes take issue with the IEA, accusing them of being an organization more committed to maintaining the established energy dynamic than telling the unvarnished truth. The IEA is an organization made up of about 150 energy experts and statisticians from 26 member countries, who since 1974 have acted in an advisory capacity on energy issues for countries both in and outside that group. Conspirators? I doubt it. In fact, their publications the last few years are noteworthy for stressing the need for accelerated development of renewable energy sources.

On the following page is another comparison that they did in conjunction with the Organization for Economic Cooperation & Development (OECD). Note that these prices figure in an 85% load factor while, as mentioned above, nuclear plants

---

[177] (IEA), "Projected Costs of Generating Electricity – 2005 Update."

are usually running at about 90% or better (thus lowering the cost of electricity), and many of the plants are expected to substantially outlive the 40-year lifespan used in these calculations. The infrequent refueling of IFRs takes only a few hours, so their load factor could very well approach 100%. Clearly the lifespan figure is extremely conservative too, as many nuclear plants are expected to last sixty years or even more. Considering that fuel and operating costs are minimal and the vast majority of the cost is sunk into construction, the real cost of generating electricity from nuclear power is astoundingly low. With IFR plants there is every reason to believe that the costs will be even lower than projected here.

	nuclear	coal	gas
Finland	2.76	3.64	-
France	2.54	3.33	3.92
Germany	2.86	3.52	4.90
Switzerland	2.88	-	4.36
Netherlands	3.58	-	6.04
Czech Rep	2.30	2.94	4.97
Slovakia	3.13	4.78	5.59
Romania	3.06	4.55	-
Japan	4.80	4.95	5.21
Korea	2.34	2.16	4.65
USA	3.01	2.71	4.67
Canada	2.60	3.11	4.00

Some comparative electricity generating cost projections for year 2010 at a 5% discount rate[178]

US 2003 cents/kWh, Discount rate 5%, 40-year lifetime, 85% load factor.

You'll notice that nuclear comes out cheaper than gas in every country, and cheaper than coal, even with these conservative parameters, in all but two countries. But is this a fair comparison? After all, there are external costs that haven't been factored in. That is precisely the thing that the European Commission wanted to examine when they

[178] Uranium Information Center (UIC), "The Economics of Nuclear Power," (Melbourne: Australian Uranium Association, Mar 2008).

launched their Externalities of Energy (ExternE)[179] study in
1991. This study was not meant to weigh the obvious costs
we dealt with here, but rather the costs to the environment,
public health, etc. It was the result of more than twenty research projects conducted over a ten-year period by researchers from all EU member states.

> The report of a major European study of the external
> costs of various fuel cycles, focusing on coal and
> nuclear, was released in mid 2001 — ExternE. It
> shows that in clear cash terms nuclear energy incurs
> about one tenth of the costs of coal. The external costs
> are defined as those actually incurred in relation to
> health and the environment and quantifiable but not
> built into the cost of the electricity. If these costs were
> in fact included, the EU price of electricity from coal
> would double and that from gas would increase 30%.
> *These are without attempting to include global warming.*
> [emphasis added]
>
> The European Commission launched the project in
> 1991 in collaboration with the US Department of Energy,
> and it was the first research project of its kind "to put
> plausible financial figures against damage resulting from
> different forms of electricity production for the entire EU."
> The methodology considers emissions, dispersion and
> ultimate impact. With nuclear energy the risk of accidents
> is factored in along with high estimates of radiological
> impacts from mine tailings (waste management and
> decommissioning being already within the cost to the
> consumer). Nuclear energy averages 0.4 €ents/kWh,
> much the same as hydro, coal is over 4.0 €ents (4.1-7.3),

---

[179] European Commission, Research Team of the Externe Project Series
(2007 [cited]; available from http://www.externe.info/team.html.

gas ranges 1.3-2.3 €ents and only wind shows up better than nuclear, at 0.1-0.2 €ents/kWh average.[180]

Bear in mind that the costs considered above relate to thermal nuclear plants, not IFRs. If we were to consider the added advantages of IFRs, most of the negatives relating to nuclear power in these calculations would be either severely reduced or eliminated altogether. The risk of accidents is reduced to almost an impossibility with no pressurized cooling systems and passive shutdown capability built in. The radiological impact of mining and milling is avoided completely since all those operations could be shut down for several hundred years while we burn up all the nuclear waste we've been reluctantly collecting all around the world and utilize depleted uranium for the rest of our needs. Given these factors, it would be surprising if the external costs of nuclear would exceed even those of wind power. Unlike wind, though, it would always be there in abundance whenever we need it.

But if those costs are so low with even thermal nukes, and nearly nil with IFRs, perhaps we should examine another point of difference between the two, the costs of spent fuel disposal that were factored into the figures we looked at earlier. Estimates of those costs vary wildly, tending to be inflated by anties and deflated by the nuclear industry. But let's try to get a feel for it.

The Independent, a noted London paper, had an article that estimated the cost of disposing of the UK's spent fuel at some £85 billion (USD$170 billion).[181] Considering the paper's liberal reputation, we can probably safely assume that the number is on the high end. On the other hand, The Yucca Mountain repository will cost an estimated $43.6 billion before it is completed

---

[180] (UIC), "The Economics of Nuclear Power." See also http://www.externe. info/team.html

[181] Jonathan Brown Steve Connor, "Tackle Nuclear Waste Disposal First, Warn Advisers," *The Independent* Jan 24, 2006.

in 2116, according to DOE budget documents. We've already spent over $8 billion on it.[182] So assuming the real numbers lie somewhere in between ('cause you *know* the DOE is hedging), that would mean that the spent fuel disposal and security costs worldwide probably run in the neighborhood of at least a couple hundred billion dollars. So how much of that could we save if we switched wholesale to IFRs?

Well first off, the costs of nuclear power generation cited above were figuring in not just the cost of waste disposal but also the cost of fuel. With IFRs the small amount of short-lived nuclear waste can be stored on-site for the entire lifetime of the plant. Disposal at the end of that time, when the plant is decommissioned, would incur a negligible cost, since the radioactive lifetime is a few hundred years, and as we've seen earlier, it will be environmentally inert.

As for the cost of the fuel, first of all we'll want to use up all of the old weapons-grade material from the military and all the nuclear waste from thermal nuclear plants. Since we'll need quite a bit of new fuel just to start up the IFRs (which will be self-sustaining thereafter), all of the nuclear waste and weapons material currently causing so much consternation can be used up as fast as we can convert it to IFR fuel. Since several countries have spent billions of dollars digging repositories for their spent fuel in consideration of having to keep it safe for tens of thousands of years, we might as well use those very expensive holes in the ground. From a standpoint of both security and economics, it would be sensible to move the spent fuel that is currently scattered across the various countries to the partially finished repositories like Yucca Mountain in the USA and its equivalents elsewhere (France, Sweden and Finland have projects in the works already). This would remove any threat of terrorists targeting spent fuel ponds, the single

---

[182] Matt Bradley, "Spent Nuclear Fuel Edges Closer to Yucca," *The Christian Science Monitor* Jul 27, 2006.

big repositories being simple to guard and quite impervious to terrorist attacks.

As for the reprocessing that would be necessary to convert spent fuel from thermal reactors into IFR fuel, that process is a bit more involved than the sort of recycling that would go on at the IFR sites in their integrated pyroprocessing facilities, each one of which will be quite modest in its capacity, sufficient to its purposes at its parent facility. The logical course of action would be to build just a few high-capacity pyroprocessing facilities capable of converting large volumes of spent thermal fuel to IFR fuel just outside the central repositories. There, conversion of the long-lived nuclear waste could be accomplished safely and expeditiously. Then as IFRs are built the fuel can be shipped to them. It only has to be done once for each IFR, since from then on they'll be able to run solely on depleted uranium that could literally be brought in by hand.

Thus the waste disposal costs and fuel costs that were factored into the economic models previously cited diminish greatly with IFRs (especially once we start fueling them solely with depleted uranium). The very conservative cost calculations that found nuclear power to be the cheapest form of electrical generation will be trumped by the even cheaper cost of power from IFRs. Both real costs, and certainly the external costs of IFRs, will surely prove superior to even the best of the modern thermal reactors.

We needn't wait for years to get an idea of the cost, either, for GE Hitachi Nuclear Energy (GEH) already designed the IFR during the Advanced Liquid Metal Reactor program and could begin building the first one as quickly as the permits could be issued. (The details vary from some of the basic descriptions I've used in this book, of course, though the essential features are the same: passive safety, metal fuel, sodium coolant at atmospheric pressure, etc.) Their original design, called the PRISM (Power Reactor Innovative Small Module) was boosted to a higher capacity and is known as the Super PRISM, or S-PRISM. An S-PRISM would be built of multiple power blocks, each containing two reactor vessels with an electric output of 380MW apiece. Each power block would feed a steam turbine producing 760MW.[183] Using a single control room, more blocks could be added even as the initial power block is operating, thus allowing power plants to come online sooner, to benefit from a common control center, and to scale up as demand increases. A typical power plant might eventually contain three to five power blocks with a total electrical output of 2 to 4GW. Alternatively, some power blocks could be used for electrical output while others utilize their thermal output directly for desalination purposes or perhaps even boron oxide recycling, a demonstration of their multi-purpose flexibility.

The word "module" in the name reveals a significant benefit of the S-PRISM, for the reactor vessels will be small enough to be fabricated in factories, then moved to the site of the power plant for installation. This is a tremendous cost and quality control advantage over nuclear plant construction in the past, where reactor vessels were constructed on-site.

All these features — simplified design, modular construction, shared systems and factory fabrication — are so far removed from previous nuclear plants as to make cost comparisons of dubious utility. In testimony before the U.S. Senate in late 2006, a GE representative presented a cost estimate for the S-PRISM of $1.3 billion per gigawatt.[184] This is well below the cost estimate of $2 billion that I use in this book, and only slightly higher than the estimates for the AP-1000 LWRs

---

[183] Allen E. Dubberley, "S-Prism Fuel Cycle Study" (paper presented at the International Congress on Advances in Nuclear Power Plants (ICAPP), Cordoba, Spain, May 4-7, 2003).

[184] "Testimony of Kelly Fletcher of GE," in *U.S. Senate Energy & Water Subcommittee*, U.S. Senate Appropriations Committee (Washington, DC: General Electric, Sep 14, 2006).

that Westinghouse is due to start building soon. In point of fact, if a massive building project of S-PRISM reactors such as the one I am proposing were to be enacted, there is every reason to believe that the cost per GW would be driven down by economies of scale and standardization. Certainly when GEH presented those cost projections they weren't imagining the implications of building thousands of identical reactors.

Those who adamantly oppose nuclear power often base much of their argument on old data rather than new technologies like the S-PRISM or the new passive safety reactors such as the Advanced Boiling Water Reactor (ABWR), the AP-1000, and the ESBWR. Cost estimates are often breezily dismissed as figments of the manufacturer's imagination. Yet there is a substantial database of real-world experience that backs up the argument that these reactor designs are indeed quite economical.

The ABWR was licensed in the late Nineties and the first two were constructed in Japan. Whereas the first build of a nuclear reactor can be expected to run over budget and face costly delays, the first ABWR at Kashiwazaki, Japan was loaded with fuel just 36.5 months after the first concrete was poured. The two reactors there are now in their fourth cycle of successful operation, with more under construction elsewhere in Japan and two others being built in Taiwan.[185]

The advantageous learning curve in any such enterprise involving standardization of design has resulted in a significant reduction in costs, so that based on real-world experience GEH can price new reactors at $1.4 billion/GW. Because of this experience, however, even more factors have already been identified that are expected to further reduce the cost to $1.2 billion/GW. These are not numbers pulled out of the air to satisfy potential customers. They demonstrate that the wildly inflated cost

---

[185] John Redding, "GE's ABWR - Key Features & an Update," *Nuclear Plant Journal* (Sep-Oct 2000).

figures tossed around by anties simply have no basis in fact when one considers these new technologies.

Both standardization and the concept of modular construction is a game-changer for the nuclear power industry. And the cost reductions seen with the ABWR will be taken to even more advanced levels by the AP-1000, the ESBWR, and the S-PRISM. All three of these designs improve upon the already impressive safety features of the ABWR with passive safety systems that further simplify their construction and thus reduce costs. Westinghouse estimates their price for an AP-1000 at a cool billion dollars per gigawatt. The ESBWR and the S-PRISM have estimated costs slightly higher, though in the event of a major commitment to nuclear power and resultant construction of dozens of reactors, those costs can be expected to at least approach the \$1 billion/GW range, just half of the conservative \$2 billion estimate I used to calculate the cost of any building spree.

With all the talk of passive safety, just how safe can we expect these reactors to be? Here we return to the science of probabilistic risk assessment (PRA) we touched on earlier. One of the salient factors of PRA that must be remembered here is that if one reactor has a one in a million chance of experiencing a certain type of accident in any given year, a million reactors could have one of their number expected to suffer such an accident every year.

At the risk of getting ahead of myself, let's do a quick calculation to figure out the PRA of two designs, the ESBWR passive safety lightwater reactor, and the S-PRISM fast reactor. Since even reprocessing all the spent LWR fuel in the world would still not allow us to provide enough fuel to load sufficient IFRs fast enough to extricate us from our dependence on coal, oil and gas, if we're serious about abandoning fossil fuels quickly we'll have to build some safe LWRs to fill in the gap until we can move to all-IFR system.

In short order we'll see just how many power plants it would take to provide all the power humanity needs if we derived all of it from nuclear power plants. Of course that won't be necessary because hydroelectric power already supplies some and renewables are growing by leaps and bounds. But we'll go by this all-nuclear assumption as a worst-case scenario, in keeping with our conservative estimates up to this point.

Let's assume that we can only build half as many IFRs as we'd like, and that to fill in the gap we have to supply half the world's energy needs with the ESBWR. Using the risk assessment figures that have been worked out for that new passive safety LWR, and dividing it by the very large number of reactors that would need to be built, we find that we could expect one core melt accident of the severity of Three Mile Island every 5,100 years or so. (Keep in mind that no one was hurt at Three Mile Island.) But since these reactors would only be a stopgap measure until we could fuel enough IFRs for a wholesale switch, we'd only have those LWRs online for about sixty years, not 5,100. The chances of a serious reactor accident are thus exceedingly remote, to say the least.

What, then, could we expect of the S-PRISM design in terms of its probable risk? After all, they have a considerably smaller output per reactor vessel than the ESBWR, and ultimately we're looking to supply (hypothetically) all the energy humanity needs. Once all the LWRs are gone, what do the probabilistic risk numbers tell us after we've adjusted them to take thousands of S-PRISM reactors into account?

As we've seen earlier, the IFR concept that is personified by the S-PRISM was developed specifically to be about as fail-safe as humanly possible. And the adjusted risk assessment numbers for even such a huge number of reactors reflect the success of the IFR concept. They reveal that we could expect a core melt accident on the order of Three Mile Island, with these thousands of reactors online, about once every 435,000 years![186]

Let's put that into perspective a little: The last ice age had glaciers gripping much of the earth's land mass about 18,000 years ago. Neanderthals died out about 30,000 years ago. *Homo heidelbergensis,* a forerunner of *homo sapiens,* was living in Great Britain about 400,000 years ago, and possibly hunting elephants with spears, or at least scavenging their carcasses for meat. According to most archeologists, the protohumans that would eventually evolve into *homo sapiens* began using controlled fire between 200,000 and 400,000 years ago.

The designers of the S-PRISM understand full well that further design changes could be incorporated to make it even safer. But really, how safe does a reactor have to be? I think if we can expect a serious accident once every several ice ages, that should be good enough for even the most paranoid among us. Of course these numbers are academic, for long before such immense spans of time have elapsed we (or whatever humans evolve into) will be traveling the stars, and cataclysms terrestrial or extraterrestrial could well make even a serious reactor accident seem trivial in comparison. We'll have long since graduated to more sophisticated energy production systems anyway. Risk assessment numbers as fantastic as this are simply indicative of an unprecedented degree of inherent safety.

So once the growing fleet of IFRs consumes all the spent fuel from LWRs, what should be done with Yucca Mountain? Will we still need it, or someplace like it, to dispose of the small amount of short-lived fission products that will issue from the IFR plants? Not really. We could cut the cost of disposal of nuclear waste to nearly nothing and save at least a couple hundred billion dollars if we simply dump it in the ocean.

---

[186] Based on GEH PRA documentation obtained by the author.

WHAT?!?! I hear you shriek. Wait a minute, hear me out before you toss your book into the fireplace. As we've seen before, the nuclear waste that we're talking about will be radioactive (beyond normal background radiation levels) for a few hundred years. When it leaves the plant it will be in the form of a glass that is designed to be inert for literally thousands of years, sealed in lead-lined stainless steel casks, the Yucca Mountain versions of which are being engineered to last at least ten thousand years. Now consider that amphorae from Roman shipwrecks that have lain for 2,000 years at the bottom of the Mediterranean Sea are still intact, having been made of simple pottery clay. We needn't go through all the trouble and expense of fabricating the 10,000-year Yucca Mountain-style casks. Instead we can simply reuse some of the casks already employed for waste storage. It can be safely assumed that those casks would last at least as long as Roman amphorae once they're lying on the sea floor. But they wouldn't even have to. A small fraction of that time would suffice. In fact, because the glass alone in direct contact with the seawater wouldn't dissolve to any appreciable degree for thousands of years, even the casks would be expendable except for purposes of shielding on the trip out to deep ocean areas where the glass could be safely jettisoned.

Sure, I realize that a few unfortunate deep-sea marine worms and their exceedingly rare buddies might end up getting toasted if they happen to wander over that way to see what just arrived from the surface — only if we dumped it raw, without the casks — but would that really be a big deal? Compare the loss of a few worms to the cost and concern of storing this stuff on the surface. Or is the apprehension of dumping it in the ocean just a matter of principle? Is all the heedless ocean dumping in the past to be conflated with this procedure? Would that be rational? I submit to you that the disposal of all the waste from the entire fleet of IFRs, be they numbering in the thousands, would most logically and harmlessly be accomplished by deep sea dumping. And when you get right down to it, we might as well not even save the casks; we can save the worms instead. After all, the casks would surely have some residual radiation inside and we wouldn't have any use for them. Since the waste from the IFRs will amount to only about 10% of the volume of spent fuel from thermal reactors, we'll have more of those casks than we know what to do with. We'll probably want to ultimately dump them empty just to get rid of them.

We should probably discuss just how much waste we're talking about, even though the vastness of the deep ocean bottom is hardly going to be affected. The entire waste output of short-lived fission products from a one-gigawatt IFR plant running for an entire year would be a little smaller than the size of a standard filing cabinet. Say we eventually have 5,000 IFR plants running worldwide once all other generating systems have been taken offline. That would annually produce a block of waste approximately 10 meters on a side, produced from the entire world's contingent of reactors for an all-electric (and boron) planet. Every few years or so we could put it all in a junk freighter that could be towed out to sea and sunk. The freighter might come to rest on a couple of marine worms. Apologies all around.

By now I hope I've made a convincing case that nuclear power, even from thermal plants, is not the economic Waterloo that environists often claim it to be. Study after study by international groups of experts using decades of actual data have shown that nuclear power is very cost competitive even today, and if you add in the external costs to the environment it comes out looking like a real bargain. Now make it even cheaper by considering the safety, proliferation resistance, free fuel and complete elimination of the waste problem offered by IFRs (unless you happen to be an exceptionally unlucky worm), and tell me what's not to like? Is there any better method of providing massive amounts of primary power to the ever more energy-hungry citizenry of the planet? At this point, if that last question doesn't seem strictly rhetorical, then you haven't been paying attention.

The Bottom Line

When we look at the cost of building nuclear plants, we see quite a disparate set of figures. Of course the numbers we have to work with pertain to thermal reactors of various types: pressurized water reactors (PWRs, of which the EPR is the latest incarnation), boiling water reactors (BWRs) (both of which are types of light water reactor [LWR]), heavy water reactors (HWRs), and gas-cooled reactors of a couple varieties that have been employed here and there, among others. The costs for a one-gigawatt plant vary from about a billion dollars (in the Czech Republic) to two billion (here in the good ol' USA with our one-off designs and legions of lawyers). In countries that have settled on standardized designs the costs approach the lower end of that spectrum.

We actually have some real-world experience in the building of commercial fast reactors. In 1972 what was then the Soviet Union built a sodium-cooled fast reactor on the shore of the Caspian Sea, in what is now west Kazakhstan. The BN-350, while capable of generating 350 MW of electricity, was instead set up for dual purposes. It produced 150 MW of electricity and the remainder of the energy was used for desalination, some 120,000 cubic meters of water per day. This was a prototype, designed to demonstrate the economic viability of such an integrated system, which it did quite successfully. Not only that, but the Soviets reprocessed the fuel in a pyroprocessing system much like the one envisioned by the Argonne project. A 1995 analysis by Argonne National Laboratory had this to say about the BN-350:

> Experience has shown that the operation and maintenance costs (reliability, availability, capacity factor) of power generation for the BN-350 plant are economically competitive with traditional (fossil-fuel or light water reactor) power plants; however, the capital cost was high

for this demonstration plant.

> ... Investigations of pyrochemical processes for fast
> reactor fuel have resulted in enough information to
> proceed with the design of a production-scale plant.[187]

Though there exists a substantial cumulative base of knowledge in regard to fast reactor design and operation, all such reactors built to date have been prototypes, and thus have incurred costs substantially higher than proven designs. When it comes to calculating costs, several companies and governments have attempted to estimate realistic costs for production model fast reactors, ranging from $1.2 billion/GW to $2.5 billion/GW. A study of proposed fast reactors of the S-PRISM type, an LMR similar to those being proposed here, calculated a capital cost of $1.3 billion/GW.[188]

Taking into consideration the economies of scale and increased efficiency developed in the course of building thousands of reactors of essentially the same model, using $2 billion/GW seems to be a more than reasonable estimate for our purposes here. Let's not forget that fuel costs are virtually eliminated, especially when we get to an all-IFR system. I realize that when we're talking about building thousands of reactors it might seem glib to speak dismissively of half a billion dollars times several thousand reactors. Nevertheless, remember the cost calculations above, which took into account a cost of up to $2 billion/GW and still found nuclear to be the low cost leader, even using other very conservative parameters. One should also consider the tremendous environmental benefits that will be an inextricable part of shutting down the fossil fuel industry. Even if it turned out that IFRs cost a bit more per kilowatt-hour than

---

[187] International Nuclear Safety Center, "Overview of Fast Reactors in Russia and the Former Soviet Union," in *Internal Document* (Argonne, IL: Argonne National Laboratory, 1995).

[188] ICAPP, "Proceedings of ICAPP '03" (paper presented at the International Congress on Advanced Nuclear Power Plants, Cordoba, Spain, May 4-7, 2003).

pulverized coal, it would be folly to choose the coal plant with all its detrimental effects and very real external costs.

When considering global deployment, the fabrication of the various components for the IFRs would be divided among companies in several countries. Everyone would get a piece of the very large pie, while competition would assure reasonable prices for all the components. $2 billion per gigawatt is assuredly erring on the side of conservatism. This would be the largest public works project in the history of the planet, by a long shot. There's simply no denying that economies of scale would apply.

Let's start by defining one IFR plant as consisting of either a single core or multiple core reactor complex, with an output of 2.5 GW at a cost of $5 billion, complete with pyroprocessing facilities and a modest underground storage area capable of holding all its waste for the lifetime of the plant, after which it could be disposed of at sea. Of course I realize that political considerations may well make ocean trench disposal unrealistic, but that's immaterial. If people want politicians to dump their tax dollars into Yucca Mountain for emotional reasons, then so be it. It'll be safe enough even there for a few hundred years, just an appalling waste of money.

Even with a program lacking the urgency that we could — and should — attach to this plan, France built six reactors a year. China is reportedly bringing a new large coal-burning power plant on line at least once a week, over fifty per year in that country alone.[189] Much of the construction of a large coal plant is virtually identical to a nuclear plant: cooling towers or other condenser systems, steam turbines, power line infrastructure, etc. Since we want to replace all fossil fuel generating plants with IFRs even though it would mean shutting down serviceable power plants, a considerable cost savings might well be realized by siting the IFRs at the

---

[189] Peter Fairley, "Part I: China's Coal Future," *MIT Technology Review* (Jan 4, 2007).

newer coal or gas plants, and splicing them into the already-built generating infrastructure there. The goal here — and an entirely realistic one — would be to build 100 IFR plants a year worldwide, starting in 2015 (giving time to test out the systems and design the plants). That would amount to 250 GW coming on line every year. Since most of the projections depicting our goals for getting global warming under control usually talk about how far along we can be by 2050, let's use that time frame. By then we would have 8,750 GW of electrical generating power online or under construction from a total of 3,500 power plants. Would we even need that much? Given that the price tag for this little building spree would run to about $17 trillion and change, are we even talking about something that's feasible here?

Let's ask an economist. How about Sir Nicholas Stern, former chief economist of the World Bank who was commissioned by Tony Blair's government to do a study of global warming and put some numbers to it. That wasn't a bad idea, sort of an attempt to take it out of the realm of scientific theory and frame it in a way that politicians could get their arms around. Sir Nicholas presented his report in October of 2006, and here, in a nutshell, is what he had to say:

Global warming is a very real threat, and unless we take serious steps to figure out how to stop what we're doing it's going to cost us an ungodly amount of money and lives. Stern's recommendation: commit 1% of global gross domestic product (GDP) per year to the task for the next fifty years to find and implement a solution. If not, break out the SPF 3000 sunscreen and head for high ground, and while you're at it you'd better pick up some weapons because with all the displaced millions there's going to be some serious fighting going on.[190]

---

[190] CNA Corporation, "National Security & the Threat of Climate Change," (Alexandria, VA: 2007).

Of course, being an economist, Stern really didn't have any solutions to the problem short of throwing money at it. Carbon trading and such made up the bulk of his recommendations, and you may remember the deservedly short shrift we gave that in Chapter Two. But his recommendation of 1% of GDP to prevent a catastrophe sounds pretty reasonable, even though it represents a staggering $500 billion a year now, projected to rise to about $650 billion by 2050. For that price, Sir Nicholas hopes that we can reduce our output of global warming gases to somewhere between 25% and 70% below current levels by mid-century. Stern estimates that the costs of extreme weather alone could reach up to 1% of world GDP per annum by the middle of the century, and will keep rising if the world continues to warm. So it seems like it's a matter of pay now or pay (more) later.

Averaging out Stern's GDP numbers, we're looking at an annual cost of about $575 billion/year to budget for our global warming battle. (Please, politicians, restrain yourselves. The first guy who uses the phrase "War on Global Warming" should find his head on a platter. But you just know it's coming, don't you?) So by 2050 Sir Richard's budget calls for spending roughly $25 trillion to reach his relatively modest GHG emission goals. You'll recall that my proposal calls for spending $8 trillion less than that amount on the construction of 8,750 GW (8.75 terawatts, or TW) of generating capacity by that same date. So will that be enough power to beat Stern's goals, at a substantial discount? If so, we'd actually be solving the problem for considerably less than Stern's price tag of 1% of global GDP. Let's not forget that whereas Stern's proposal assumes a drain on the global economy, these power plants pay for themselves. So the $8 trillion that it looks like we're saving on the face of it really amounts to a whole lot better deal.

At the present time global electrical production amounts to about 2.3 TW, with nuclear power supplying about 16% of that amount. Here's a chart from the IEA's publication *World Energy Outlook 2004* to help us crunch the numbers. As you can see, 7% of total global energy supply (not just electrical production) comes from nuclear plants with a current combined capacity of 368 GW. Even though there is a big push to bring the portion of renewables dramatically upward, we shouldn't forget that energy demand is predicted to double by 2050, so to be very conservative we'll assume, as stated previously, that the

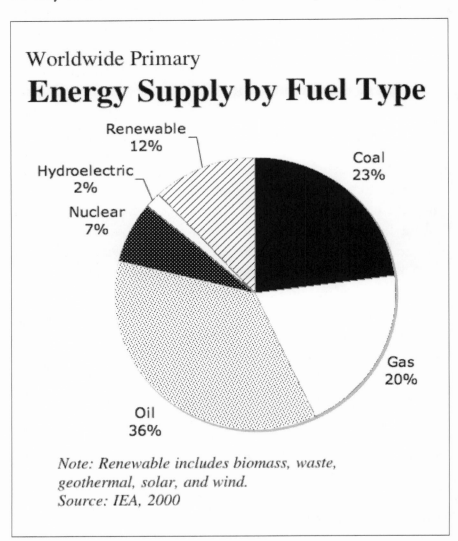

Worldwide Primary
Energy Supply by Fuel Type

Renewable 12%
Hydroelectric 2%
Nuclear 7%
Coal 23%
Gas 20%
Oil 36%

Note: Renewable includes biomass, waste, geothermal, solar, and wind.
Source: IEA, 2000

percentage of renewables and hydropower will remain the same. Plasma converters also have the potential to supply a considerable amount of electricity, by some estimates as much as 5% of demand in the USA from municipal solid waste alone. Nevertheless, since there are so many uses for the syngas they produce, I will completely discount any contribution of plasma converters and assume that any electricity they contribute just provides more padding for our consistently conservative estimates.

Coal, oil, and gas comprise fully 79% of energy supply, all of which we intend to replace with IFR nuclear power. The oil referred to in this pie chart refers mostly to its use in vehicles. But remember we're replacing that with IFRs using boron as an energy carrier. If the plasma conversion works for boron oxide recycling as I speculated earlier — meaning that garbage would provide the energy needed for the boron recycling — that would make a big difference in the amount of energy we'd be looking to replace with IFRs, allowing us to build a lot fewer power plants. But we'll stay on the safe side here and assume that we'll have to rely on the IFR power to recycle the boron for our vehicle fleets. Plus let's not forget that we also want to replace that 7% that is being produced by thermal nuclear plants as they come to the end of their useful lives, which will be the case for many (though not all) of them by 2050.

Thus we have to provide about 12 times the amount currently provided by nuclear power to completely replace those energy supplies of today. But with energy demand doubling by 2050, we'll have to be able to provide not just 12 but fully 24 times as much power as nuclear does today. That would come to approximately 9 TW, which is just a shade over what our proposed building project would provide, at a cost of about two trillion dollars per terawatt. For purposes of estimation and in consideration of the consistently conservative assumptions built in throughout these calculations, 100 power plants a year can be considered entirely realistic as a target to meet our mid-century energy needs.

Just how realistic is it to think we can build 100 nuclear plants per year? Remember that France built up to six per year during their conversion to nuclear, so let's look at Gross Domestic Product (GDP) as a guide to what a given country can financially bear for such a project, keeping in mind that France proceeded without the sense of urgency that the world today should certainly be ready to muster. There are six countries with higher GDPs than France, all of whom already possess the technology to build fast reactors: USA, China, Japan, India (they're building one now), Germany, and the United Kingdom. Add Canada and Russia (which already has one running and is planning more), then tally up the GDP of these eight countries. At the rate of 6 plants per year with France's GDP, these countries alone could afford to build about 117 IFRs per year, even without any greater urgency than the French brought to bear on their road to energy independence. And come on, you know that using "urgency" and "French" in the same sentence is pushing the envelope.[191]

Of course there are plenty of imponderables in an estimate like this. Part of it would depend on what kind of efficiency we can get out of our boronmobiles, and what level of efficiency the boron recycling plants would be able to achieve. If plasma converters can handle boron oxide recycling as theorized, it would eliminate that energy demand altogether. Since boron has considerably greater energy density and burns much hotter than gasoline, efficiency in vehicles shouldn't present a problem. In the event that plasma reduction doesn't fly for boron oxide and the heat for recycling has to be generated thermally and electrically, IFRs could handle it. Since the fuel for the recycling plants is essentially free (coming from our free-fuel IFRs) and the boron is also free after the first tank is purchased except for the minor costs of transport and recycling and a modest profit

---

[191] Full disclosure: I lived in France for two years, so I do have some personal experience with which to support such a jest.

for the retailer, efficiency isn't as important as one might suppose. Of course that would impact the number of recycling plants necessary to process the boron. The replacement of natural gas for heating buildings and for industrial processes will also be subject to a conversion/efficiency ratio as gas is replaced by electricity. If governments see fit to subsidize energy-efficient heating and cooling systems like geoexchange heat pumps, the conversion rates should be quite favorable. Bear in mind that the very low cost of boron and the low cost of electricity would make it possible for modest taxes on either or both of those commodities to fund programs such as subsidized energy-efficient conversions, without undue pain to society at large.

Improvements in energy efficiency will very likely make a major impact as well, as can be seen by the prior example of California. While per-capita demand for electricity in the rest of the United States was increasing by fifty percent, California's per capita use stayed flat for thirty years. Since the 1970's, California's energy-efficiency standards have reduced electricity consumption by the equivalent of the output of more than 20 average power plants.[192] (New York, too, is doing great in this regard.)[193] Imagine the savings if such policies were widely applied around the world.

I live in California myself, and I can testify that the sort of programs implemented here have not been painful in the least. New technologies are being developed all the time to make appliances and other electrical devices even more efficient. California has a long history of creating technological hurdles for manufacturers to jump over. Because of this the state has led the way on everything from catalytic converters to power limits on electronic charging units, the so-called "vampires" used for

---

[192] Felicity Barringer, "California, Taking Big Gamble, Tries to Curb Greenhouse Gases," *New York Times* Sep 15, 2006.

[193] Statemaster.com, "Total Electricity Consumption (Per Capita) by State," (National Priorities Project Database, 2001).

everything from cell phones to toothbrushes that account for up to 10% of a modern home's electrical use.

It's not hard to foresee the day when enlightened public policy would seek to apply energy standards to all equipment that uses electricity or natural gas (until we manage to replace all natural gas). Like CAFE standards in the automobile industry, government could continually ratchet up its demands to push energy-saving technologies to new frontiers of efficiency.[194] Applied globally, such advances could easily obviate the building of dozens, probably even hundreds of power plants. Clearly this isn't science fiction. California programs applied just to the rest of the states could cut electricity demand by about a third. It's a question of political will, not technology. The technology just keeps getting better. Lest we lean too heavily on the Stern Report, it might be illuminating to look at some other energy projections to see how they stack up with the IFR/boron plan. After all, perhaps Sir Nicholas is just an incorrigible pessimist, or despite his status as a world-renowned economist maybe he forgot to carry a couple of zeros. The IEA came out with a report to celebrate their 30th anniversary that had a lot to say about our energy future.[195] Let's look at their numbers for what they see as investment needs in the energy sector. Bear in mind that these figures aren't envisioning any goals regarding global warming or the many other problems addressed by our IFR proposal. These are just what they see coming as a matter of course if we stick to business as usual:

- More than $16 trillion, or $550 billion a year, needs to be invested in energy-supply infrastructure worldwide over the

---

[194] Yes, I know the USA pretty much abandoned the idea of tightening CAFE standards in recent years, another of its shameful policies of environmental neglect.

[195] IEA, "30 Key Energy Trends in the IEA & Worldwide," (Paris: International Energy Agency, 2001).

three decades to 2030, an amount equal to 1% of projected gross domestic product.

- The electricity sector alone will need to spend almost $10 trillion to meet a projected doubling of world electricity demand, accounting for 60% of total energy investment. If the investments in the oil, gas and coal industries that are needed to supply fuel to power stations are included, this share reaches more than 70%. Transmission and distribution will account for more than power production.
- Total investments in the oil and gas sectors will each amount to more than $3 trillion, or around 19% of global energy investment. Coal investment will be almost $400 billion, or 2%.
- For the energy sector as a whole, 51% of investment in production will be simply to replace existing and future capacity. The rest will be needed to meet the increase in demand.

Don't say I didn't warn you about big number vertigo. Now let's examine each of those points and see how they compare to our energy plan:

> *More than $16 trillion, or $550 billion a year, needs to be invested in energy-supply infrastructure worldwide over the three decades to 2030, an amount equal to 1% of projected gross domestic product.*

You'll note that this pertains only to the year 2030, not 2050 as I've proposed. This is calculated as money that will be spent regardless of efforts to solve the problems we've tackled with our plan which would cost about $17 trillion instead of this study's $16 trillion — pretty much a wash with estimates like these — but ours would cover twenty more years to boot! It looks like the IEA's estimates of what Sir Nicholas wants to spend through 2050 look surprisingly close to each other. $17 trillion would be a bargain and we'd solve a host of serious problems that today have nearly everyone in gloom and doom mode.

> *The electricity sector alone will need to spend almost $10 trillion to meet a projected doubling of world electricity demand, accounting for 60% of total energy investment. If the investments in the oil, gas and coal industries that are needed to supply fuel to power stations are included, this share reaches more than 70%. Transmission and distribution will account for more than power production.*

Okay, here's a factor that we didn't take into account with our $17 trillion figure: the transmission and distribution infrastructure. According to the graph that accompanied these projections, 54% of the electricity cost will be needed for transmission and distribution. That's no small amount when we're talking about trillions of dollars. We can, however, dispense with the oil, gas, and coal portions they warn about here, since they won't be involved. Still, that bill comes to some $5.4 trillion dollars, and that's only through the year 2030. If we project that same rate of spending out through 2050, that brings our $17 trillion project cost to a whopping $26 trillion. Everybody take a deep breath. Remember, the first part of this section said we'd have to spend $16 trillion through just 2030, so we're still in the ballpark of the do-nothing scenario costs.

But a do-nothing scenario entails huge costs to society. As mentioned before, coal soot alone is killing 24,000 Americans (and far, far more Chinese) prematurely every year and costing the country $167 billion dollars annually. In a business-as-usual future, coal has been predicted to be used even more, but even if we just stayed at our present grim levels until 2050, that $167 billion/year would add up to over seven *trillion* dollars. From soot! And that's just in the USA. Estimates of premature deaths from coal soot in China have run in the staggering range of 400,000 people per year or more. As to the financial costs to that environmentally benighted nation, we can only guess. But take that seven trillion from the USA's coal soot costs to 2050 and just imagine what the global costs are if China alone is so much worse off. How about Eastern Europe? The other nations under the Asian Brown Cloud? Clearly the global societal costs of coal-fired power plants from now until mid-century must easily surpass twenty trillion dollars (Do the math). Though we're in the midst of comparisons here, the situation is really beyond comparison. Eliminating coal in favor of IFR development is simply the sane choice, even if we don't take global warming, nuclear proliferation, resource wars and all the rest into account. We certainly should, of course.

Just looking at the money we'd save in the USA by replacing our coal plants, that $167 billion per year would buy 33 IFRs of 2.5 GWe each per year. Just with the money the USA would save going forward by replacing its coal plants with IFRs, this one country could finance a full third of the entire global cost of this energy revolution single-handedly. From soot savings! It would take roughly 13 years to replace all the coal plants in the USA with their equivalent generating capacity in IFRs at the rate of 33 IFRs per year. If the entire country adopted the California model, electrical demand could stabilize enough to ward off any substantial increase in electrical demand during that time. If not, it will take a bit longer, but at the very least in 20 years all the coal plants could be gone. At that point we'd have an extra $167 billion per year that we're now pouring down a soot hole—a hole that's also swallowing about 24,000 of our citizens every year.

Lest we forget some other changes that will be wrought by this plan, by then the USA will also have extricated itself from the oil quagmire of the Middle East, where it's presently spending about $120 billion per year on yet another money hole — war in Iraq — that's consuming both resources and lives. Imagine the good uses to which over a quarter trillion dollars a year could be applied. Those savings alone could finance over half the world's conversion to IFRs.

> *Total investments in the oil and gas sectors will each amount to more than $3 trillion, or around 19% of global energy investment. Coal investment will be almost $400 billion, or 2%.*

These are costs that can be completely dismissed, since these fossil fuel industries would be shutting down. The coal industry will cease to exist, since even smelters will be able to convert to electricity. This may sound horrific from an employment standpoint until you realize that since WWII the coal industry shed 80% of its work force as automation took over. This will be but the culmination of a trend that is inevitable and already largely accomplished. As Jeff Goodell points out in his excellent book, *Big Coal,* the coal miner population is rapidly aging to the point where most of them will soon be retiring. Unlike in past eras, for the most part their children have not followed them into the mines. It is already a dying industry.

> *For the energy sector as a whole, 51% of investment in production will be simply to replace existing and future capacity. The rest will be needed to meet the increase in demand.*

Every one of the IFRs will be brand spankin' new. Being modular units, the lifespan of the plants will likely be greatly extended, since the parts will be able to be replaced as long as the structures themselves are sound. Designing the structures with a hundred year lifespan in mind should not be difficult, nor should it be too hard to build them with access for module replacement in mind. Such foresight in the design process can accommodate even the replacement of the reactor core itself, as well as all the constituent parts of the reprocessing facilities. About the only piece too large for replacement as a single unit will be the stainless steel vat containing the sodium pool in which the reactor core is immersed. However, since the experience at Argonne Labs showed the corrosion of that component to be essentially zero over a period of thirty years of operation, it is doubtful that the vat (and its enclosing secondary backup vat) would ever need replacement. As for meeting the increase in electricity demand mentioned in the IEA report, we've already accounted for that in our plan.

Not being used to bandying about numbers in the trillions, I have to say that nine trillion dollars worth of distribution and transmission systems sounds awfully pricey, but then again there's a lot of work to be done in developing countries, many of which have virtually no transmission lines. I strongly suspect that the costs can be reined in, however, through the use of nuclear batteries to accomplish distributed generation and eliminate the need for thousands of miles of high-tension power lines. Nevertheless, we'll use the projections in the IEA study, projected forward to 2050 as I've done above. So our project has grown to a total of $26 trillion through the year 2050.

That's a few trillion dollars less (and remember our estimates have been consistently conservative throughout) than what the IEA says we'll be spending through just 2050 if we extrapolate their figures that far out. Presumably they're anticipating that by 2030 there'll be a good number of currently operating power plants still online, though aging and requiring replacement before 2050. Our plan, though, will have nothing but brand new squeaky-clean IFRs humming along, solving the problems of nuclear waste disposal, air pollution, environmental degradation from mining and drilling, oil wars, water wars, and nuclear proliferation. I know it's hard to think that anything costing $26 trillion is a bargain, but in this case it definitely is.

Despite the fact that our IFR/boron/plasma plan adds up to less than the IEA and Stern numbers, the uncertainties associated with such projections might make it seem not all that much better. (Well, they sure do to me, but I'm addressing the cynics in the crowd here.) Bear in mind that the IEA's projections were only meant to cover normal energy use and did not take into account solving the many problems addressed by our plan, nor any of the attendant costs that such solutions would incur. This is money that they envision as being spent in any event, global warming and other issues notwithstanding. So the price tag of our program is actually trillions less than what we would be actually saving when we consider all the added benefits which the IEA simply didn't address in their calculations.

The savings in relation to the Stern Report's projections are likewise understated, for Sir Nicholas seems to have ignored the extra costs of power distribution and transmission, a little detail that substantially increased the cost of our plan. If you add those costs into his projections so we're comparing apples to apples, it balloons Stern's figures to nearly $35 trillion through 2050, fully nine trillion dollars more than our plan. Everett Dirksen's quip that led off this chapter takes on an exponential resonance when billions have been transformed into trillions. We are indeed talkin' real money here.

The estimates for upgrades to the distribution and transmission infrastructure that we added in here might end up costing considerably less, though, which will only improve the picture. There's been a new type of power line developed that can carry at least twice the current of existing high-tension lines.[196] It has

---

[196] Reuters, "BC Transmission Corporation Chooses 3M's Aluminum Matrix Conductor for Two Segments of Vancouver Island Transmission Reinforcement Project," (May 14, 2008).

been tested in the field for years and is now being installed in several countries.

This technological advance is especially welcome when we consider that societies all over the world will be changing to all-electric households and industries. Building new towers and establishing new power corridors would be terribly expensive. Fortunately it won't be necessary. With these new Aluminum Conductor Composite Reinforced (ACCR) power lines, more than twice as much current can be carried on existing towers. All the aluminum from the existing lines can be recycled and incorporated into the new lines, which are likewise predominantly comprised of aluminum.

Even accounting for unforeseen cost variables in our calculation of the IFR project's costs, there is a HUGE amount of wiggle room for this plan. Indeed, by any logical estimation it should represent a savings of at least ten trillion dollars over anybody's alternatives, while reaching goals of planetary health that others can't even bring themselves to hope for in their most optimistic projections.

The Stern report has been criticized for greatly exaggerating the cost of climate change and inflating the chunk of GDP it proposes for addressing it.[197] One can argue the arcane points of the social discount rate and other such socioeconomic factors till doomsday, however. It matters not in terms of our plan, for as one can see from a comparison with the IEA projections, our energy plan would be spending less than even a business-as-usual approach to future energy needs. The IEA figures aren't even considering global warming except insofar as political pressures will encourage renewable energy development, a minor consideration in their calculations in any event. By any measure, even for those who reject Stern's analysis wholesale, our plan is a bargain. One way or the other, trillions are sure

---

[197] Hal R. Varian, "Recalculating the Costs of Global Climate Change," *New York Times* Dec 14, 2006.

to be spent meeting the world's future energy needs, whether or not you factor in global warming and regardless of the price you put on its potential costs. The salient question in any case is how the money is going to be spent. If we can spend less of it than anyone has so far projected and cut the Gordian knot of global warming in the bargain, where's the downside?

Before we conclude this comparison, it's been suggested that the true GHG cost of nuclear plants should, in order to be honest, include the GHGs that are emitted in the course of the construction of the plants themselves. Fair enough, but if we consider that power plants of one kind or another will inevitably be built to satisfy demand, there's nothing inherent in the building of an IFR that would appreciably increase the emissions compared to building any other kind of large power plant. In point of fact, by the time we begin the actual construction of IFRs in 2015 (according to this plan), earthmoving equipment and the other vehicles involved could all be running on boron, with zero emissions. As for the cement involved, yes, cement production does indeed produce carbon dioxide, a little over 2% of total global emissions of this greenhouse gas (bear in mind that cement is used all over the world for construction, much more so than in the United States, where its relative contribution is less than 1%[198]). It is possible that building IFRs might well utilize somewhat more cement per GW of power capacity than building power plants fueled by coal or gas. Even if that turns out to be the case, the fraction of the already small fraction involved hardly seems worth mentioning when we consider the tremendous advantages of the IFRs over all other types of power plants. I defy anyone who would reject this plan based on the amount of GHGs emitted during construction to support their resistance with even a shred of logic.[199]

---

[198] U.S. EPA, "Sources of GHGs in the U.S.," (Pew Center on Global Climate Change, 1998).

[199] I'm thinking of you, Ms. Caldicott.

It must be recognized, though, that GHGs arise from a variety of sources, as pointed up by the cement example. The IEA report cited above estimates that by 2030 about half the global production of GHGs will come from electrical power generation and about 25% from vehicles. If our plan didn't even touch industrial or residential sources, then, we'd still have cut our emissions by 75% just by implementing it, assuming that many homes and businesses would still be heated with gas and that steel smelters would still be burning coke. Yet that is quite unrealistic. Since ample and quite economical electricity would be available for these uses, the conversion to an all-electric society would proceed quite naturally, slashing GHG emissions in the process. Despite the protestations of those horrified by social engineering, it would be dead easy (and alas, probably necessary) to push the conversion along by the simple expedient of taxing fossil fuels to the point where the last of the holdouts would be forced to acknowledge the new reality. Some people get upset with revolutions; what can I tell you? Get over it. Besides, we'll be able to make carbon-neutral gasoline from garbage-derived syngas for the real diehards and antique car buffs. It'll probably even be cheaper than gas is today.

So we've seen that both the IEA and the Stern Report's projections are considerably more pessimistic than what we could accomplish with the IFR/boron/plasma plan. By 2050 we will have spent about $9 trillion less than Stern recommended while completely replacing all non-renewable power supplies, and instead of getting halfway to zero GHG emissions we'd be virtually all the way there (cement and a few other industrial processes notwithstanding). Sir Nicholas and I differ greatly, then, and here's another reason why: "Throw a lot of money at it and hope it goes away" is not really a plan.

But we do have a plan, and the money isn't just being thrown down a hole. All the nuclear plants and boron systems will pay for themselves through the sale of electricity and boron, even considering that the prices for those commodities will be quite low by today's standards, especially the boron in comparison to our gas prices. Or will they? Haven't we forgotten the profits that have to be raked off into the pockets of the plutocrats?

CHAPTER NINE

Cui Bono?

Follow the money.
— Deep Throat, *All the President's Men*

T HE DRIVERS OF almost any country in the world, no matter what their nationality, have always been able to readily agree on one thing: Americans are spoiled rotten when it comes to what they pay for gas. Not quite so true anymore, though. When gas prices topped three dollars a gallon in 2006 there was real pain at the pump for a lot of Americans. Having gone on an SUV buying binge since the early nineties, they now were seeing the chickens coming home to roost.

So you can imagine the reception that the oil companies got when they started lamenting that they had so much cash that they really didn't know what to do with it. The word chutzpah doesn't really do them justice. Callous, maybe? Insufferable? Definitely. Criminal? Well, at least criminally insensitive. Sudden strident calls for a windfall profits tax fell upon deaf ears in Congress. The oil giants' money had been talking too loudly for way too long for the lawmakers to hear the cries of outrage now.[200]

---

[200] Note to non-U.S. readers who may be unfamiliar with America: The U.S. government operates under a legalized bribery system. This sort of thing is par for the course.

One would be hard-pressed to imagine just how any oil executive could possibly think that whining about having too much cash wouldn't have repercussions while Americans were being bled at the gas pumps. But then it's also difficult to imagine just how much money we're talking about here. In 2005 and 2006, ExxonMobil was consistently reporting quarterly profits in excess of $10 billion, about $110 million a day — more profit than any company in history. It's a testament to the political power of oil — and to the moral bankruptcy of the D.C. establishment — that at that same time the conservative-led Congress (give credit where credit is due) was forking over to Exxon and its fellow oil giants an additional $4 billion in tax breaks and subsidies.[201] Talk of a windfall profits tax to ease the sting that Americans were feeling at the pump never made it beyond the talking stage. The same situation continued virtually without a hiccup after the Democratic party took "control" (sort of) of Congress in 2006. The only difference seems to be that Americans are paying even more for gas, and the quarterly profits of the oil giants are more obscene than ever.

It's not just in the modern day that oil companies have been powerful enough to pursue their rapacious ways with impunity. Standard Oil, started by the Rockefellers and their cronies in 1870, grew so audacious and economically brutal that despite herculean efforts at skirting the laws meant to rein them in the company was finally forced to break up into 34 separate companies in 1911. Ever since then the oil companies have been slowly reassembling those fragments back into giant entities, like some sort of diabolical self-repairing robot in a science fiction movie. Thus we find the oil industry today controlled in the main by six major players: Exxon/Mobil, Conoco/Phillips, Chevron, BP, Shell, and Total, the first four of which were originally the fragments of the forced breakup of Standard Oil.

---

[201] Tyson Slocum, "Big Oil Can Afford to Forgo Tax Breaks — but Renewable Energy Can't," in *Public Citizen* (Feb 27, 2008).

The political power of these companies extends worldwide, and until the formation of OPEC in the Sixties and the flexing of its muscles in the early Seventies, foreign producers were controlled almost like colonies of the oil companies. They have been involved with wars, coups, uprisings and/or assassinations in nearly all the oil-producing nations of the world. Today, with two oilmen at the head of the U.S. government (well, one oilman and one failed oilman), it's not surprising that we find U.S. troops in Iraq, a nation with some of the largest proven oil reserves in the world. Oh, the Bush administration has stridently denied that this war is about oil, despite their Freudian appellation of the invasion as Operation Iraqi Liberation (later hurriedly changed to Operation Iraqi Freedom when the press picked up on the unfortunately revealing acronym).

Of all the industries in the world there is probably none that has been the object of as much conspiracy mongering as the oil industry. Nor is there probably any industry that has deserved it more. The utter dependence of the world on transportation, relying almost entirely on oil, means that those who control the oil supply have their hand on the world's jugular. Nobody likes to be at the mercy of anyone, much less the tender mercies of an entity with such a rich history of deception, manipulation, greed, and even war.

But let's be fair. Greed is what corporations are designed for. When the head of BP, Sir John Browne, admitted that the threat of global warming was real in 1998 and began making sounds about corporate responsibility he was roundly taken to task for it. There were complaints that his environmental consciousness-raising might be contrary to the interests of BP's shareholders. Just like Henry Ford was castigated nearly a century earlier for not trying to squeeze enough profit out of his company to keep the stockholders satisfied, Browne suffered similar slings and arrows. Nevertheless he managed to diversify BP into alternative energy technologies while his fellow oil CEOs were still in denial.

Global warming has gotten so alarming, however, that even the other oil companies have gotten involved in alternative fuels. Talk of peak oil and the burgeoning demand in China and other rapidly developing nations makes it a certainty that alternatives must be found. Most of the oil companies have hydrogen R&D programs, which suit them well on many levels. For one thing, there are so many technological hurdles to be leapt that the realization of the vaunted hydrogen economy is still comfortably distant. In this respect hydrogen research functions quite effectively as a greenwashing strategy, which the oil industry can point to as proof of their commitment to the environment even as they continue with business as usual. Indeed, it would be naïve in the extreme to think that the oilmen haven't been instrumental in such egregious anti-environmental moves as the refusal to improve fuel efficiency standards for American vehicles. One of the most outrageous moves of all, though, was the Bush administration's orchestration of tax policy to allow a $2,000 tax rebate to owners of the 60-mpg Prius while at the same time giving owners of the 11-mpg Humvee a $25,000 tax break.[202]

Investing in hydrogen research and putting their political muscle behind its eventual adoption as the primary fuel of the future has important implications for the oil giants. If oil is destined to lose its primacy as the fuel of choice for transportation, how better to maintain the lucrative control over energy supplies than to pursue a technology that virtually guarantees a complex and expensive infrastructure? What sort of companies will have the financial clout to even attempt to control such a colossus? Trillions of dollars are at stake. A highly volatile fuel like hydrogen would require even more special equipment and safeguards than gasoline requires today. It's doubtful that oil company executives care all that much about where

---

[202] Jeffrey Ball & Karen Lundegaard, "Loophole Gives SUV Buyers a Tax Break," *Salt Lake Tribune* Dec 20, 2002.

our energy comes from. It's their business to control it no matter the source.

During this book's gestation period I had many conversations with people in all walks of life, for the basic outline of the idea had occurred to me several years ago. The most common reaction I got once someone was informed of the IFR and boron technologies was first pleasant surprise, then resignation. I was repeatedly warned that unless I could engineer a way to give the energy giants control over the system, the concept was doomed to failure. They are simply too powerful to defy.

So who deserves a piece of the pie? Should the coal companies get in on the action even though coal will be obsolete? Do the oil companies have any more right than the coal companies to control of the world's energy supplies which will have nothing whatsoever to do with oil? They're all getting involved with biofuels too, but when biofuels can be simply generated from garbage that will be a moot point. Does the public really want to turn over control of their future energy supplies to a cabal that has taken every possible opportunity to gouge them for the past hundred years?

And what of the private utility companies? They'll surely be jockeying for position at the trough as well. With trillions of dollars at stake they're not just going to go gentle into that goodnight. But electric utility companies have hardly shown themselves to be faithful stewards of the public trust either. And like the oil companies, they have long since insinuated themselves into the corrupt culture of Washington in order to pursue their understandably selfish ends at the expense of their customers.

Utility companies, like oil companies, have a long history of influencing legislation to their advantage, as can be expected. Thus it was considered perfectly acceptable for them to make a healthy profit for years while the government was keeping them tightly regulated. The policy of price caps under which they operated had been put in place more to allow them to make a profit than to protect their customers from exorbitant charges, for their costs of the power they sold was quite low, often considerably lower than what they were charging. Yet the system worked well for all concerned, resulting in healthy utility companies and still reasonable electric rates with a very high level of reliability. Pretty much a win/win situation.

Then the deregulation genie started to peek out of the lamp. It started in Great Britain under Maggie Thatcher in 1990 and was later copied by a few other countries and some U.S. states, most infamously California. Basically what the system did was to split the utility companies' functionality into two separate groups: generators and distributors. The former own the plants, the latter own the grid. Once that bit of sleight of hand is accomplished, the stage is set for the public to be fleeced.

The generators offer bids for their electricity that they plan to generate the next day, and the distributors purchase their power based on the resulting competition. It's the sort of thing that's music to the ears of diehard capitalists. Of course in any such situation there's an opportunity for entirely useless but exceedingly avaricious traders to get their foot in the door, rather akin to day traders on the stock market, who buy and sell and, if successful, make scads of money while contributing nothing (except stock volatility) to the companies whose stock they're trading.

The story of California's "energy crisis" of 2000-2001 illustrates just what can happen when a necessity of life like electricity is allowed to be wholly exposed to the mercies of untrammeled market forces. Once California decided to go down the road to deregulation — with the gleeful encouragement of the state's main utility companies, Pacific Gas & Electric (PG&E) and Southern California Edison (SCE) — the utilities sold off their generating plants at immense profits, which they duly pocketed. Never mind that those plants had been purchased at their customers' expense. They'd spent tens of millions of dollars to "convince" lawmakers to go ahead with deregulation. Now they were going to start the harvest.

In the few years leading up to the "crisis," PG&E and SCE drained nearly ten billion dollars out of California, sending it back to their parent companies.[203] Their generating plants had been purchased by out-of-state companies who would be complicit in the energy trading fiasco that ensued. Unfortunately for those California utility companies, though, deregulation wasn't *quite* complete, for the legislation had left price caps in place beyond which they could not charge their customers. Those caps were plenty high for the utilities to make a hefty profit, as they had for years. But when all hell broke loose the caps were to be their undoing.

Out-of-state traders, most notoriously but certainly not exclusively the infamous Enron Corporation, began gaming the system so outrageously as to make their British counterparts look like pikers. (Actually, Enron and the others had already been active in Great Britain, where they'd honed their skills at ripping off the public. California was to be their larcenous utopia.) When demand was spiking during the long hot summer, the perfectly serviceable power plants that the utilities had recently sold suddenly started going offline for "maintenance." Electricity prices which had pretty accurately reflected the generating costs when hovering around 5¢/kWh soared as high as 52¢. But the utilities weren't allowed to pass the entirety of those costs on to their consumers because of the price caps. The monster they had helped create had come back to bite them with a vengeance.

Appeals to the incoming Bush administration fell on deaf ears. This was not surprising, since the majority of the energy traders participating in the larceny were based in Texas and

---

[203] John Dunbar & Robert Moore, "California Utilities' Donations Shed Light on Blackout Crisis," in *Center for Public Integrity* (May 30, 2001).

had longstanding ties to the Bush dynasty. "Kenny Boy" Lay, as Bush liked to call him, the CEO of Enron, had been a friend and generous contributor to the Bushes for years. The Federal Energy Regulatory Commission (FERC) had been packed with regulators sympathetic to the bandits who were now busy bankrupting the utilities and driving California citizens to distraction. Rolling blackouts became a regular occurrence as monthly electric bills soared to the allowable limits.

But those limits were not nearly high enough for the utility companies to pay their bills to the energy traders, whose continued manipulations were reaching new heights of creativity with each passing day. By the end of May 2001, SCE and PG&E were asking their customers and the state's taxpayers to bail them out with over $13 billion to pay off their debts to the energy wholesalers. Just one month prior to that plea, PG&E had filed for bankruptcy, *one day* after awarding their top managers some $50 million in bonuses.[204]

WARNING: IRONY OVERLOAD! Prior to this debacle, as mentioned above, PG&E had sent nearly $5 billion from the sale of their California power plants sluicing back to their parent company. Across the country in Bethesda, Maryland, one of the offspring of that parent corporation, National Energy Group, had been experiencing a dizzying rise in their fortunes due in no small part to infusions of capital from said parent company. In a mere ten years NEG was more valuable than its 96-year-old sister company in California, PG&E, that had so faithfully contributed to its meteoric rise. By April of 2001, the month PG&E filed for the aforementioned bankruptcy, NEG was the nation's third largest power trader. The source of much of their profits? You guessed it: California![205]

---

[204] Ibid.

[205] Richard A. Oppel & Laura M. Holson, "While a Utility May Be Failing, Its Owner Is Not," *New York Times* Apr 30, 2001.

How is it that the free market principles with which many are so enamoured failed so spectacularly when applied to the electricity market? The incomparable Greg Palast offered this insightful perspective:

> I first came to Britain in 1996, to help the incoming Labor government try to fix the nation's new — but already broken — electricity market. It didn't work. Year after year, the fixes failed, as they will fail in California and other states that think they can design a deregulated system. There is no fix: Free markets in electricity go berserk because they aren't really markets, aren't free and can't be. Electricity isn't like a dozen bagels; it can't be frozen, stored or trucked where needed. And while you can skip your daily bagel, homes and industry will not do without their daily electricity.
>
> As a result, deregulation is never really deregulation but an unhappy mish-mash of rules belatedly chasing runaway prices generated by each week's new trading game. To salvage their imploding market, the California power pool's economists busily craft one wacky fix after another — "Intra-zonal Congestion Management," "Price Volatility Limit Mechanisms" and more, which tumble out of their bureaucracies like circus clowns from a Volkswagen. A delicious irony is that "deregulation" has produced an explosion of shifting regulations and new bureaucracies dwarfing California's old regulatory system.
>
> Market fundamentalists say the solution to half-baked deregulation is full deregulation, with no rules at all. That's frightening. As former World Bank economist Joe Stiglitz said to me the other day, these theorists are like medieval bloodletters. If a dose of their free-market medicine doesn't cure the patient, they call for applying more leeches.[206]

---

[206] Gregory Palast, "Some Power Trip," *Washington Post* Jan 28, 2001.

While the story of California's ludicrously disastrous experience with energy deregulation seems unlikely to be outdone anytime soon, it is most definitely not rare as an example of runaway corporate greed and its consequences for electricity consumers, aka all of us. Mismanagement, waste, neglect, and worse have more examples among America's utilities than we'd like to acknowledge. Though we can only imagine what scenarios play out with private utility companies in other nations, there is one hard and fast rule that has proven itself over and over again in every energy industry, be it oil or gas or electricity: If the opportunity to gouge the customer is available, eventually the customer will get gouged. Repeatedly, if possible.

> The problem with this model is that the buyer is nowhere so free as the seller. A person, an office, a city cannot simply do without electricity for a week if the price is too high, any more than they can do without oxygen. It is the absolute dependence of the buyer that makes this a very unequal exchange — unless the distribution is regulated, or publicly controlled, or both. Instead, what California established when it deregulated the industry in 1996 was a system that maximized the buyer's vulnerability, forcing utilities to buy their power on a daily basis on the spot market. That's not a crisis of the supply and demand for natural resources. That's a crisis of a marketplace that gives all power to the seller.[207]

Energy, be it automobile fuel, heating fuel, or electricity, is necessarily prone to such susceptibilities, at least in industrialized countries, because for all intents and purposes it is one of life's necessities. The ground is fertile for ex-

---

[207] Harold Meyerson, "Power to (and from) the People!," *L.A. Weekly* Feb 2, 2001.

ploitation: huge amounts of money, a captive audience, and the virtual impossibility to either forgo the product's use or have a meaningful option to avail oneself of competitors' products. Oh sure, there's some competition out there when you go to fill up your car with gas, but the increasing monopolization of the oil industry has made real competition naught but a fantasy.

One of the ironies of the California debacle was the striking fact that in the midst of the crisis there were some cities that seemed immune to the problems. Los Angeles and Sacramento both continued with business as usual — no blackouts, no rate hikes. The problem, of course, was not an actual shortage of electricity but a robbery in progress. These cities were unaffected because they had their own municipal utilities that weren't held hostage like their hapless corporate comrades. Many Californians who saw the lights burning brightly in those cities, especially noticeable when surrounding areas were shrouded in darkness from blackouts, could be forgiven for wondering where the system had gone wrong.

The fact of the matter is that publicly owned and operated utilities have had a long and successful history in many of the American states. Nowhere is this more apparent than in Nebraska, where the state's entire electrical system is a public trust. But it wasn't all roses getting to that point, for the clarion call of private ownership was sounded there too. Nebraska, though, like many other states, had the disadvantage of having low population densities spread over a large agricultural area. Thus the pressure from utility companies to push for private ownership was not as intense as in more densely populated regions, for no private company wanted to face the costly task of building a power grid in a state where they'd have very few customers.

The situation was ripe for a different vision, and that was where Senator George Norris of Nebraska stepped in. During the 20s and 30s Norris worked tirelessly to promote publicly owned electrical systems primarily in rural areas that, like his home state, suffered a lack of infrastructure due to the private utilities being uninterested from a cost/benefit standpoint. He was the prime mover behind the Tennessee Valley Authority, which still provides inexpensive hydroelectric power to some six states along the Tennessee River. He fostered the creation of the Rural Electrification Administration, which provided power to farmers across the country through federally sponsored cooperatives. Despite vehement opposition by private utility companies, Senator Norris persisted in bringing his public power vision to his entire state. Castigated as a socialist in a time when that word carried considerable stigma (as it still does today in the USA), he succeeded in not only selflessly serving his state's best interests but in laying the groundwork for a multitude of public power systems which persist to this day.

Today there are over 2,000 government-operated systems across the country, with a third of them — like Sacramento and Los Angeles — having their own generating capacity. Altogether, the nonprofit electrical sector — publicly owned utilities plus private, member-owned cooperatives — services 26% of American consumers.[208] Condemned by the private utility companies as socialistic and thus somehow inherently illegitimate, these nonprofit systems consistently provide reliable power more cheaply and reliably than the corporations that deride them. It's not surprising that the deviously mislabeled "socialists" have considerably better results for their customers.

In 1948, Carleton L. Nau, then executive director of the American Public Power Association, outlined the unique advantages the absence of private shareholders gave to publicly owned utilities at the municipal level. These included the substitution of community well-being for

---

[208] Wayne O'Leary, "Electricity Illuminates the Ghost of George Norris," *The Progressive Populist* Jan 2006.

the profit motive as an operating ethic, the ability to dedicate earnings directly to plant improvements and quick debt retirement, the freedom to apply revenue surpluses toward lower rates and better service, and (because of exemption from federal taxation) significantly lower operating expenses.[209]

Over half a century later those arguments have lost none of their validity. Indeed, they've been proven true countless times since then, the California fiasco being just the latest and most glaring example. Deregulation simply isn't an issue because there is nothing to deregulate. Yet despite the stellar track record of public ownership, the big utility companies and free market true believers routinely trot out that old bogeyman of socialism and wave it threateningly at American citizens, hoping to activate a visceral revulsion to the very concept of socialism, which found its greatest potency during the McCarthy era.

It's just jingoistic nonsense, of course, since customers pay based on consumption just like those served by private utilities. But it's a very convenient sort of fiction for those who would seek to run roughshod over the interests of their fellow citizens in pursuit of the almighty dollar. Almost every proposal to initiate government programs supportive of the people's welfare has been met originally with warnings of socialism or even communism. It's as if by hanging a tag of socialism on such endeavors, any serious discussion of such ideas can be identified with Marxism and thus placed beyond the pale of the discourse of 'true Americans.'

Such a narrowing of public debate serves only the interests of those who would seek to exploit the majority of Americans. The demagogues who utilize such spurious methods to limit

---

[209] Ibid.

legitimate deliberation have been altogether too successful, relying as they do on equating unrestrained capitalism with the core values of America. The infamous and, alas, recently resurrected Newt Gingrich was once quoted as saying, "The purpose of American government is to strengthen American companies in the world market." Now isn't that a lofty goal to inspire his fellow Americans? Presumably the enhancement of democracy, freedom, and justice are meant to take a back seat to the goals of corporate America. Those who elevate the market to the top of the list of things they wish to keep free too often put the average American far down that list.

Make no mistake about it, America is a socialist country. Indeed, it would be quite impossible to find any country that isn't, since the term simply connotes some degree of publicly funded benefits which accrue to all a society's members. Without that there would be no society. While many Americans look askance at the socialism they see so prevalent in the EU, often those same Americans refuse to recognize the same phenomenon right outside their own doors. The fact is that our country has a lot of socialism that we take for granted in our everyday lives. Transportation infrastructure, public education, national defense, and all manner of public facilities and services are accepted socialistic arrangements that we not only count on but also expect and appreciate. So it's not a question of socialism versus capitalism, as some demagogues would have you believe. Rather, it's a question of where we decide to draw the lines. To buy into the notion that the very concept of socialism is somehow antidemocratic and unpatriotic is utter claptrap. Such reactionary biases only serve to limit rational debate at a time when it's desperately needed.

When contemplating an entirely new energy structure not only for the United States but for the entire world, it would be foolish beyond measure to discard the possibility of publicly owned and operated systems. Indeed, there are almost irresistibly compelling reasons to consider public ownership as being far preferable to private ownership, not the least of which is the fact that the energy revolution proposed herein is based primarily on nuclear power.

Sure, private utility companies operate nuclear power plants all over the United States and some other countries as well. But let's take a look at the life cycle of a nuclear plant in the USA and see just what the utility company brings to the table. For starters, they ask for and receive generous tax breaks and usually other subsidies to help defray the considerable capital costs of construction. Then once the plant is up and running, they are shielded from economic responsibility for unlikely but potentially ruinous accidents by the Price-Anderson Nuclear Industries Indemnity Act, a piece of legislation that allows the utilities to act as an insurance pool for each other with the assurance that the federal government will pick up the tab for anything that goes over its limit. During its operating lifetime the utility is responsible for operation and maintenance, under the not always sufficiently watchful eye of the Nuclear Regulatory Commission (NRC). It's also responsible for continually skimming off a profit for its shareholders — every corporation's prime directive. At the end of the plant's operational lifetime, the federal government is once more expected to help shoulder the cost of decommissioning, and if you don't believe that the taxpayers are going to get stuck with the disposal and security costs for the nuclear waste in the decades and centuries ahead, then perhaps I could interest you in some swampland in Florida. Oh, and the R&D that made it possible to build these plants in the first place? Government labs, funded by the taxpayers.

There are clearly way too many things wrong with this system, but one of the worst is something that many people never consider. It's called the Bathtub Curve. This is a well-recognized engineering principle charting the probability of breakdowns plotted against time of operation, ramping down steeply from the beginning when a physical plant is new. The high level at the outset reflects the increased probability of early accidents ("infant mortality") from design or equipment flaws. Then the line runs horizontally for most of the life of the plant, indicating a low probability of accidents during the plant's operating lifetime once it's made it past the early higher risk period. But then as the plant nears the end of its expected lifetime the odds of an accident rise dramatically (the other end of the bathtub) as equipment has aged and maintenance or inspections may be neglected.

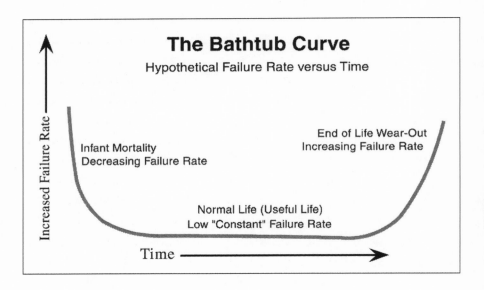

This is where private ownership has the greatest tendency to exacerbate the most negative possibilities of the bathtub curve, for as a power plant ages (or any type of industrial plant, for that matter) there is a great temptation for plant owners to skimp on maintenance and operator training while they try to wring the last dollars out of their investment. In most types of industries such behavior might result in limited accidents or possibly even the deaths of workers, but obviously in a nuclear power plant the risks carry considerably more gravity. If the NRC is really in bed with the people they regulate, as many allege, then the oversight which is their *raison d'être* will be little more than perfunctory at the very time when it's most required to be rigorous. Since the NRC is funded by the very power plants which they inspect, and since closed power plants don't generate funds for it, there is a logical incentive for the NRC to keep plants open and possibly even look the other way when potential problems surface.

Such neglect by both the owner and the NRC seems to have been in evidence with the Davis-Besse nuclear plant outside Toledo, Ohio. For up to four years, FirstEnergy Corporation's workers there (they're the private utility that runs D.B.) had ignored boric acid dripping onto the top of the six-inch-thick carbon steel lid of the reactor vessel from cracked nozzles. Despite the fact that the NRC had found such cracks in nozzles at other plants similar to D.B. and advised the utilities to inspect them, even the NRC's own inspections failed to notice the problem. The utility's inspectors did find nozzle cracks (just above a growing cavity on the lid of the reactor) and reported it to the NRC in July of 2001. Due to that and a poor safety record at Davis-Besse, the NRC ordered the reactor shut down for an inspection — no later than the end of 2001! To make matters worse, FirstEnergy strong-armed the NRC, arguing that their increased attention justified a delay until their next scheduled plant shutdown in February of 2002. Despite substantial evidence that FirstEnergy's vaunted extra attention seemed to have meant nothing in the past, the NRC assented to the delay.

By the time they got around to checking the reactor, the boric acid had eaten completely through the entire 6 inches of the reactor vessel lid, and all that was preventing a loss of coolant accident was a 3/16" stainless steel liner that by this time was bulging into the milk jug-size cavity due to the internal pressure of 2,500 psi. Having narrowly escaped a serious accident, the plant was shut down and, while they were at it, FirstEnergy repaired a sump pump system used for backup cooling in case of just such an accident. It turns out that a Los Alamos Laboratory study of these pumps, used in at least 70 reactors around the nation, reported an estimated one in three chance of pump failure in the event of an accident if not fixed by 2007. The NRC's recommendation in 2003 after receiving the report from Los Alamos: fix the sump problem by 2008.[210] Ultimately "the NRC admitted it failed to properly police the plant and had ignored numerous warnings; a survey of NRC employees found many concerned 'that the NRC is becoming influenced by private industry,' and that there is "a compromise of the [agency's] safety culture.'"[211]

These are but a sampling of a disturbingly lengthy list of the NRC's negligence. Though nuclear plant containment buildings are designed specifically to contain the radioactivity in the event of even a severe accident, clearly the task of the NRC is to prevent those worst-case scenarios. Just as clearly, they have fallen down on the job. According to a 2003 report by the NRC's inspector general and the Government Accountability Office, 47 percent of NRC employees don't feel comfortable raising safety issues.[212] But the NRC's main responsibility is to detect safety issues! Clearly the NRC should be subjected to serious and constant oversight to make sure they're doing their jobs (which is supposed to be serious and constant oversight), which they have been startlingly blasé about attending to in too many cases.

After the Davis-Besse incident, the NRC admitted they'd been negligent. On January 20, 2006, FirstEnergy acknowledged

---

[210] David Lochbaum, "Regulatory Malpractice: The Nrc's Handling of the Pwr Containment Sump Problem," (Union of Concerned Scientists, 2003).

[211] Matt Bivens, "Two-Bullet Roulette," *The Nation* Sep 10, 2003.

[212] Lewis, "Green to the Core? — Part 1."

a cover-up of serious safety violations by former workers in the near-accident, accepting a plea bargain with the U.S. Department of Justice in lieu of possible federal criminal prosecution. But this and the many other well-documented failings of the NRC point to a glaring need for an overhaul of a system that cannot afford to be unreliable. Like far too many regulatory agencies, all signs point to a too-cozy relationship between regulator and regulatee.

The passive safety features of the IFR systems would have made such accidents as Three Mile Island, Chernobyl, and Davis-Besse's near-accident a physical impossibility. Yet when proposing a vast deployment of even the safest nuclear power plants, common sense tells us that all possible precautions should be taken to safeguard both the integrity of the plants and the security of the fuel supply. When privately owned utilities are responsible for maintenance, they can be counted on to cut corners to save money, especially at the end of a reactor's lifetime when such cost-cutting is most tempting. If the watchdog of the U.S. nuclear industry can't be trusted to police the utilities, how confident will we feel about every other nation's regulatory watchdogs?

The training of power plant workers has also been the responsibility of private utilities, and has hardly had a stellar record. This sort of problem is not confined to U.S. utility companies. In 1999, two inexperienced workers at a uranium enrichment facility in Japan were killed when they decided to bypass safety protocols and mix some uranium in a large tank to save time. The flash of radiation when their misstep resulted in a criticality burst was an entirely preventable accident arising from carelessness and a lack of training, according to the Kansai Occupational Safety and Health Center:

> Though Japan has one of the most developed economies, companies like JCO save money by not training staff

properly. This is the case in this disaster... Proper training would have educated them about the dangers of working with nuclear material. Instead, the company decided minimal instruction was sufficient, and showed them how to mix the material in much the same way as a bricklayer would make up a small amount of concrete in a bucket. The result of the penny-pinching policy for two of these unfortunates was death.[213]

The implementation of a global energy plan requires some fresh thinking in order to provide oversight on a worldwide level. Elimination of the profit motive would be a great place to start. Looking at the record of publicly owned utilities in the United States, which serve fully 26% of American consumers with an enviable record of dependability and rates averaging 18% lower than private utilities, provides a strong argument in favor of extending the nonprofit model to a global scale. There would be many compelling advantages to such a plan, and few if any drawbacks. Indeed, unless one considers the inability of utility companies to skim off profits as a negative feature, there would seem to be no drawbacks whatsoever.

This is hardly a radical concept. Indeed, it is already a reality on a national level in France, where their AREVA national nuclear power agency oversees all aspects of their nuclear industry, from mining, power plant construction, training, reprocessing, and every other detail up to and including waste disposal. The only obstacle to copying their system and implementing it worldwide is political. They have clearly demonstrated its effectiveness.

---

[213] Kansai Occupational Safety & Health Centre, "Tokaimura Nuclear Accident from an Occupational Safety and Health Viewpoint," (Japan Occupational Safety and Health Resource Centre, Apr-Jun 2001).

Let's examine the features of this proposed nonprofit global energy consortium and how it will work. We'll call it, henceforth, the Global Rescue Energy Alliance Trust (GREAT). The international negotiations and hard choices required to create such a system will be formidable, requiring policies that will cut harshly against the corporate and political grain — more in some countries than others. But nobody ever said that implementing a plan to save the planet was going to be a bed of roses. In reality, though, we'll see that aside from the impossibility of placating the greediest power mongers (in both senses of the phrase), the advantages of such a system would be overwhelmingly positive for the rest of us.

Corporatist true believers (free market ideologues) will undoubtedly argue that GREAT is a matter of ideology, and its supporters will surely be tarred as socialists or even communists in the inevitable efforts to discredit this proposal. But GREAT is not a matter of ideology, it's a matter of sanity. Just as the world lived under the threat of nuclear annihilation during the long tense years of the cold war, so we will continue to live under the threat of nuclear terrorism until we recognize the fact — not the opinion — that the only way we can ever hope to remove the threat of nuclear proliferation and nuclear terrorism is to put the entire nuclear fuel cycle under strict international control. This perforce requires us to end the era of private utility companies' involvement in nuclear power.

As newclear power assumes its role as the dominant energy source of the future, the only recourse for private utilities will be in renewable technologies that contribute to the overall energy supply system. Given that IFRs and the existing thermal reactors will likely supply the vast majority of power at least in the near term, it stands to reason that the overall energy infrastructure and administration will fall under the purview of GREAT, making electrical generation and distribution a de facto near-socialized system. (Since usage will still determine users' costs, it would not be a socialized system per se, but more akin to a cooperative. But what's in a word?) If wind and solar power are practical alternatives to nuclear, as their proponents maintain, then there will be plenty of room for investment by private sector energy companies, though given the history of manipulation of energy markets it would be prudent to limit the generating capacity of any one company along the lines of PUHCA.[214]

Subsidization of any energy source will be unnecessary once the IFRs are up and running. The customers will provide a steady income stream. If it does, indeed, prove to be the case that solar and/or wind power are truly competitive with newclear on a level playing field, then we'll see an upsurge of renewable energy production in the hands of private companies and individuals. As long as the system has regulatory limits, capitalism in the power arena will thrive. There is, of course, the very real possibility (I would say probability) that newclear power will be considerably cheaper than either wind or solar. In that case they will fail to be any more than niche suppliers, for there would be no market for high-priced power under a unified and unsubsidized system. As for hydropower, it is doubtful that many more dams will be built, considering the environmental and social impacts that have hamstrung every modern hydro project and will likely continue to do so, at least in western Europe and North America.

---

[214] The Public Utilities Holding Company Act, a U.S. law mandating the regulation of electrical utilities. It was repealed in 2006, opening the floodgates for abuse that could make California's energy "crisis" look like a walk in the park. Americans, you are duly forewarned.

CHAPTER TEN

How Great is GREAT?

Electricity is really different from everything else.
It cannot be stored, it cannot be seen, and we cannot do without it...
It is a public good that must be protected from private abuse.
— S. David Freeman,
Chair of the California Power Authority, 2001

To anyone living in the United States, it is a foregone conclusion that a move to essentially nationalize, much less *internationalize,* our energy system would be nothing short of revolutionary. And it will be sure to inspire the sort of fervor on both sides of the debate that characterizes revolutions of an often bloodier sort. But this one should be bloodless, for we can see it coming and clearly discern the underlying motives as being in humanity's best interests.

Many nations, perhaps even the majority of nations, would readily recognize the benefits of GREAT, the Global Rescue Energy Alliance Trust. The very definition of the word "Trust" tells the tale: a fiduciary relationship in which one entity holds the title to property for the benefit of another — in this case, for humanity at large. For the French, as proudly independent as any people, it would be a considerably smaller step to embrace GREAT than for many other countries. But the many ramifications of GREAT would create an uproar in some segments of French society too. For the system as proposed would drive Total, one of the six oil giants and a major player on the French stage, the same place it would drive all the other fossil fuel companies throughout the world: either out of business entirely or into side industries that would leave them mere shadows of their former selves. But this is the nature of technological and economic evolution, reminiscent of the plight of the buggy whip manufacturers at the dawn of the automobile age. Fossil fuel companies are the 21st century's equivalent of buggy whip companies. Really BIG and POWERFUL buggy whip companies, to be sure, but every bit as obsolete not too many years hence.

Change or die. That's the law of the corporate jungle in such times of shifting realities. BP's purchase of solar panel manufacturing plants points to a way that the oil giants can still salvage some vestige of their old power, but it's almost a given that they will never again enjoy the clout that they wield today. The world is moving on. If oil companies begin to invest some of their obscene bankrolls in alternative energy systems that are truly clean (the plasma converter industry being one obvious path), they could survive and even prosper. If they don't do it others surely will, and in time the remnants of the once great oil companies won't have enough political and economic mojo to fight off the upstarts. Oil is for dinosaurs, appropriately enough considering where it comes from. Evolve or die. Just don't expect anybody to shed any tears.

There has been much speculation and doomsaying lately about peak oil, the observation (refuted by some, passionately asserted by others) that we've just about reached the peak of our oil extraction limits. This barrier, combined with burgeoning global demand, could inexorably lead to ever-higher prices and future resource wars. The fact that the United States and Great Britain are currently bogged down in a war in Iraq lends a special relevancy to such arguments. There is simply no way that these two western nations would be embroiled in that war if not for the stranglehold of world oil supplies.

GREAT and the technological revolution that it represents would make such wars a regrettable memory, provided we could provide ample power to all nations on an equitable basis. Certainly the resources are there to do so. The construction of a hundred or so IFR plants per year is most definitely within the realm of possibility. And the amount of fuel already on hand is so vast that our only problem would be using it all up before fusion power elbows IFRs from the stage — if that ever happens. Take a look at this rather unusual graph:

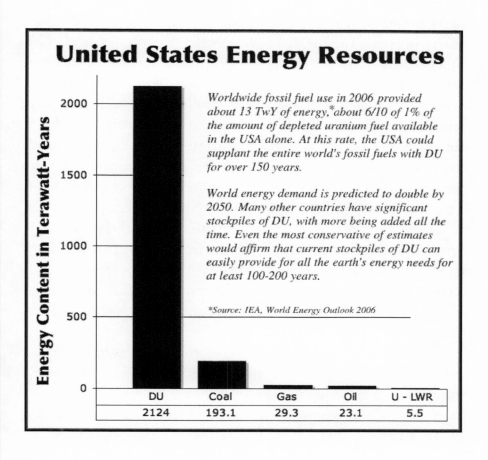

United States Energy Resources

Worldwide fossil fuel use in 2006 provided about 13 TwY of energy, about 6/10 of 1% of the amount of depleted uranium fuel available in the USA alone. At this rate, the USA could supplant the entire world's fossil fuels with DU for over 150 years.

World energy demand is predicted to double by 2050. Many other countries have significant stockpiles of DU, with more being added all the time. Even the most conservative of estimates would affirm that current stockpiles of DU can easily provide for all the earth's energy needs for at least 100-200 years.

*Source: IEA, World Energy Outlook 2006

Energy Content in Terawatt-Years

DU	Coal	Gas	Oil	U - LWR
2124	193.1	29.3	23.1	5.5

The depleted uranium bar on the left represents the available supply of already-mined and milled uranium by the time current reactors reach the end of their life cycles (most of them will do so within the next 20 years). Coal, considered by some the answer to the USA's energy future, looks positively puny, less than a tenth the height of the fast reactor bar. What this graph amply demonstrates is that once all the thermal reactors have reached the end of their life spans we could immediately shut down every uranium mine on the planet and run only IFRs for hundreds of years, worldwide.

Those who find little difficulty comprehending the concept of a kilowatt-hour will nevertheless probably do a double take at a measurement in terawatt-years, much less the way it would more properly be written in scientific terms as 2.124 *petawatt*-years. Whereas a kilowatt-hour denotes the energy expended in one hour by 1000 watts of power (burning ten 100-watt light bulbs, for example, for an hour), a terawatt-year denotes the energy expended by burning a trillion watts for a year. You may recall from Chapter Seven that the expected energy needs of the entire planet in 2050 necessary to replace all fossil fuels and thermal nuclear plants would come to approximately 9 TWy/year (plus the minimum projected contributions of renewables). That would represent a doubling of current energy needs, as predicted by the IEA and others. If we assume a further doubling by the end of the century and a smooth ramping up from this point (at this time nuclear power supplies only .368 TW), a tricky calculation (why didn't I pay more attention in calculus class?) yields an approximate energy need from now until 2100 of about 900 TWy.

If the United States alone were to offer to supply the *entire world's* non-renewable energy needs solely with IFRs burning the free fuel we've already accumulated, we would use less than half of our available supply by the end of the century. By that time demographic experts expect the population to have stabilized and reversed, which would be a blessing since we'll have about ten billion souls here at the peak, assuming we don't destroy ourselves before that. Would we all be using more energy per capita or less by then? Hard to say, but besides the remaining prodigious supply of fuel that we'd still have left in the USA, don't forget that the above calculations didn't even consider all the uranium that's sitting around all over the rest of the world right now. By any reckoning, we already have sufficient uranium mined and milled to power the earth for many centuries.

One would think that we could simply close down uranium mining entirely, which will certainly be the case once IFRs have taken over. But there are still 435 thermal reactors operating around the world, and they will all continue to need new fuel until the end of their operational lifetimes, that point arriving for most of them before mid-century. While it might be tempting to think that we could continue to fuel them from reprocessed spent fuel, the fact is that if we're going to start up a hundred IFRs per year we'll use up all the spent fuel in the world just on their initial fueling needs, and still be wishing we had more.

Here we come to the problem of deploying nuclear plants in non-club nations. We have recently seen the entry of North Korea into the nuclear club, a development that nobody in their right mind (and that would seem to exclude Kim Jong Il) feels very good about. Iran's flirtation with uranium enrichment has the dogs of war straining at their leashes. Since the nuclear battery concept is still in the theory phase (though surely feasible) and perhaps not very practical on a massive basis at least in the near term, what possible way could we expect to site nuclear reactors in non-club countries that could, at any time in the future, become politically destabilized?

We can get a clue by looking at the internationally accepted concept of embassies, a concept alluded to a few chapters back.

An embassy's structures and properties, even though situated in a foreign land, are still recognized by international agreement as a small plot of foreign soil, under the sole jurisdiction of the particular embassy's home country. This same concept could be expanded to create international energy embassies in countries around the world, under the sole jurisdiction of GREAT. Transport of fuel into these sites would be done under its supervision. All energy facilities would be subject to random inspection by GREAT's own inspectors.

The question of the operating crews is one of the best parts of this system. GREAT would establish a corps of highly trained IFR plant operators made up of individuals from all over the world. Only these crews would staff the critical managerial and technical positions at the energy embassies. As suggested earlier, lest some scurrilous nation might try to coax a team of operators to divert some fuel away from the site (difficult though that would be from a logistical level, since it will always be highly radioactive), the crews would be subject to reassignment at any time. Since each power plant would be the same as all the others (GREAT having settled on a standard design), moving the operators from one plant to another, from one country to another, would not present an expertise gap. Indeed, it would present an added safety opportunity.

The last thing a new hire wants when coming into a job is to be blamed for a predecessor's mistakes. It would be standard procedure for every incoming crew to conduct a thorough inspection of the plant into which they're transferring. Not only that protocol but also their natural inclination to maintain their good reputations (the CYA factor) would assure their diligence. Thus the operating crews themselves would take the place of the NRC or its equivalent. This protocol could be further enhanced by roving teams of GREAT inspectors whose only job would be yet another layer of oversight, with unannounced inspections any time, any place.

Political unrest, however, can never be fully anticipated, and there would always be the possibility of a rogue ruler or rebel group trying to take over an energy embassy by force. In the extremely improbable event that they could gain entry (a futile endeavor, provided that sensible design is employed), it would be nigh impossible to cause any type of nuclear accident within a short period of time because of the inherent safety features of the IFR design. And there would be no weapons-capable material on the site at all. But clearly such a takeover could not be tolerated. Thus an indispensable arm of GREAT would be a strike force on full standby, trained specifically to retake an energy embassy in hostile hands. In point of fact, if such a force ever had to be called upon it would be to defend one of the current generation of reactors, for the IFRs could easily be designed to be virtually impregnable.

Nevertheless, for the most insecure among us let's imagine a hypothetical attempt to take over an IFR energy embassy. The collateral damage involved in any such operation would be neg-ligible, as each facility would be surrounded by an uninhabited perimeter, a no-man's-land, in effect. If a security force were to be instituted by international agreement, their preauthorization to strike in the event of a hostile takeover would give serious pause to any armed group that might even contemplate such a move. What, after all, would they hope to accomplish? They could not cause a nuclear explosion, which would be impos-sible. They would have great difficulty even moving any fuel out of the plant, since the pyroprocessing facilities keep the mixes radioactively very hot. Nothing of any use as a weapon (except in a crude non-nuclear dirty bomb) would be available, and even that would be virtually impossible to utilize because of its highly radioactive nature. Plus the certainty of quick and lethal retaliation would be a formidable deterrent.

The international community certainly has the military ca-pability to maintain a positively scary GREAT strike force. With every power plant under constant communication via satellite, including site cameras both inside and out, an attack on a plant could barely be underway before the force was en route. Indeed, few military upheavals happen in a vacuum. In the event that political unrest is in the air, the strike force could be whisked to a nearby location on standby, or even stationed temporarily at the energy embassy itself in the hot zone. It would be a simple matter to design the reactor complex as a hardened target virtually impossible to penetrate, even without a strike force in the vicinity. Blast doors for all ingress could even be operated by remote control from a central command post outside the country. With satellite communication and modern detection and remote weapons systems, a properly designed IFR complex would be untouchable by even the most sophisticated terrorist group. A single security person would be more than adequate.

Every nation not already in the nuclear club would be required to sign on to GREAT's standard international agreement, willingly surrendering sovereignty of the energy embassies and accepting, in advance, the political consequences of even attempting to usurp that authority. Why would a country submit to such demands? For the many advantages of cheap, clean unlimited energy, of course. What country, it may more logically be asked, would refuse such an opportunity?

A system like this, far from contributing to global tensions, would have precisely the opposite salutary effect. GREAT would usher in a new era of peaceful cooperation, allowing virtually every nation to be energy independent. Energy is the common denominator of every society, no matter the political ideology. Every nation needs it to develop, as Paul O'Neill, former U.S. Treasury Secretary, so passionately stated in Ron Suskind's book *The Price of Loyalty*. After a trip through Africa with the always-surprising Bono as his incongruous companion, O'Neill came back to the USA pointing out that if the industrialized countries are actually serious about helping the so-called developing countries make progress, then we must help them attain two basic needs: safe water supplies and ample electricity.

Those who may have suspected until this chapter that I was possibly a shill for the nuclear power industry will now surely realize that that could not possibly be the case. I dare say that the private utility companies that own and operate the nuclear plants in the USA are not going to consider me a popular fellow at all. They can get in line with the fossil fuel companies. I can take it; I'm a big boy. As FDR once said, "Judge me by the enemies I've made."

The Boron Factor

Let us not forget that our system stands on three legs, not one. What of the boron component? The logic of internationalization of electrical generation certainly has a fairly undeniable logic in a nuclear/boron society, but it might be argued (vociferously, if I know the oil companies) that the boron business should remain in the control of private companies even if the nuclear component does not. There are again, however, a lot more arguments against that position — and precious few for it — if we're going to consider the benefit of humanity and not those of our would-be corporate energy masters.

The relationship between the boron recycling plants and the IFRs will necessarily be a very close one. The boron recycling method used today requires higher temperatures ($700°C$) than IFRs have been designed for (about $550°$). A temperature boost using electricity can be utilized to bridge the gap. That said, high-temperature reactors are in development with an eye toward thermal hydrogen manufacture (unnecessary in our world of boronmobiles), but these would utilize fuel types which have the distinct and deal-breaking disadvantage of producing non-recyclable waste. This is not to say there might not be one of two developments in the future: a high-temp reactor with fully recyclable fuel and the same safety factor and other advantages as the IFRs, or a different type of boron recycling method requiring temperatures no higher than the current operating temperatures of IFRs. (The plasma conversion option could also conceivably work.) In either case, dedicated recycling plants could then be developed that directly utilize the thermal energy of the reactors, thus improving the efficiency factor of the IFR-to-boron cycle. But if nuclear power is used in any such capacity for boron recycling, we would definitely want the recycling plants to be under the direct control of GREAT.

Since our nuclear fuel is free, any possible energy penalties of using IFRs for recycling are quite acceptable, despite the possible loss of efficiency. But there is a compelling reason to keep the boron system under the control of GREAT in one great consumer-centric energy system. This is where boron truly shines, for it would act as the great buffer, maximizing the efficiency of the entire energy structure.

One of the problems with generating electricity is that you can't store it all that easily. Converting it from one form to another and back again entails quite unacceptable losses. Another big problem is that demand is necessarily sporadic. Since power plants cost a lot of money to build, nobody wants to build too much capacity into the system knowing that much of the generating potential will be idled a lot of the time. The problem is only compounded when we begin to add in solar and wind power, for both suffer from the fickleness of nature's whims. Wind is as flighty as, well, the wind. Solar is more predictable (but those cloudy days don't help), though it obviously peaks in the early afternoon whether you like it or not, while residential demand tends to peak in the evening when people come home from work. This is especially critical during hot weather, when millions of air conditioners kick in at full blast around five o'clock, just about the time the sun is getting low in the sky.

The cost of IFRs will be nothing to sneeze at, even taking mass production into account. We don't want those plants sitting idle or running at half power. This is where the synergy of boron recycling to electrical generation can pay tremendous dividends and maximize efficiency of the total energy picture. For boron recycling plants need not run at full capacity all the time. They can run at whatever rate they can draw power. All they have to be able to do is to keep enough recycled boron available to meet local demand.

Almost everyone's had the experience of using rechargeable batteries, which can be very handy except when they start to get old and refuse to hold their charge. Any electricity storage system would have to be able to avoid that problem, and boron fills the bill perfectly because it's inert. Its potential energy today will be the same next week, next month, or next year. Thus it can act like a giant rechargeable battery to soak up excess electricity whenever it's available. When electricity demand rises, the boron recycling plants would just throttle back and produce less boron. In extraordinary circumstances they could even shut down for a while altogether, though in an integrated energy system a balance would inescapably be found to maximize both the electrical generation and boron recycling systems. Thus the grids would be provided with ample power in any contingency without the costly necessity of building needless overcapacity into the system. Wind and solar contributions would fit in seamlessly, fully integrated into the energy symbiosis, while the power plants would be able to run at full power virtually around the clock. Hydroelectric plants, of course, are fully adjustable, and reducing their flow in times of low electricity demand would only leave more water in the reservoirs for later use.

Vast outdoor storage yards would be no problem, or large grain silo-type repositories (in the event that powdered or pelletized boron is found to be preferable to spools or some other configuration). As we transition to the energy world of the future and oil refineries begin to close down, it would be reasonable to buy the land on which the refineries currently sit and convert those sites to boron recycling plants. Not only would it ease the transition for the shrinking oil companies, but there would be an almost irresistible poetic justice to it. For the people living in the vicinity of refineries have suffered the smell and pollution of the refineries for years. Now their new neighbor, the boron recycling plant, would be environmentally benign. No more smells, no more smoke. In fact, since the boron recycling process would release pure oxygen into the air (unless the plasma systems work as theorized in Chapter 7 — sorry I haven't had the opportunity to test that yet), the closest neighborhoods would actually benefit from living in a slightly oxygen-enriched environment. The effect would be greatest at night, when the boron recycling is proceeding at the plant's full capacity, while the neighbors are snug in their beds. Not exactly a hyperbaric chamber, but it wouldn't be surprising if the people in such neighborhoods ended up being healthier than average. The irony is so thick you can cut it with a plasma torch.

With such a cleanly synergistic system, any increases in power generation from solar or wind power would be utilized to their maximum advantage, while the IFR plants would also experience peak utilization. Temporary imbalances would be minor and easily compensated for by either shipping extra energy in or out to neighboring regions via the power grid or moving boron into nearby areas by truck or rail, which is quite inexpensive since boron isn't volatile and can be cheaply moved by common carrier.

It would seem obvious, then, that keeping the boron recycling under the same publicly owned GREAT system would be far superior to putting it in the hands of private companies. Imagine what would happen if Exxon Boron owned the recycling facilities instead. For one thing prices would be higher, that's a given. It's built in to the corporate model; they need to make a profit for their shareholders. But if they were building the recycling plants, they would want them to be producing at maximum capacity around the clock. Throttling back to compensate for electricity peak demand would be taking money out of their shareholders' pockets. Especially in exigencies like heat waves and cold snaps their output would necessarily drop, and who can doubt that they would use that as an excuse to raise the price of boron? Can you hear them whining now?

What, after all, would private companies bring to the table on either the nuclear plant side or the boron recycling side? Two things most certainly: higher prices overall and as much price volatility as they could manipulate into the mix. Keeping the entire system under the GREAT umbrella, on the other hand, would give us maximum efficiency and the lowest possible cost. Price volatility would be completely eliminated from the system, since electrical generation capacity would be utilized to the fullest extent and boron prices would remain stable as a rock. That is no small matter, for fuel price volatility affects our entire economy. When fuel prices rise, truckers and rail operators have to compensate by raising their prices for everything they carry, and by everything I mean just that: everything. Thus fuel volatility adds to economic destabilization on every level.

Why not eliminate it entirely if we have the chance? Because we don't want to be called socialists? Because our "free market" true believers insist that pure capitalism (which doesn't exist anywhere on earth, really) is the be-all and the end-all of the American system and that without it our rugged independence and indefatigable initiative would cause our nation to wither? Please! There is a long and proud history of publicly owned energy systems in America and around the world. The arguments in favor of such a system for energy (and for health care, for that matter, but that's another can of worms) are far stronger than the arguments for private ownership.

Let's look at the oil industry, for example. The production of oil and its manifold products is a complex business involving substantial economic risks. Just finding the oil deposits, in all their varying forms and values, is a great and inescapably expensive challenge. Drilling for it everywhere from islands in the Arctic Ocean to the bottom of the North Sea requires a degree of creativity and innovation that is quite amazing, and which pretty much has to evolve in the field. The refining process too involves incessant fine-tuning by chemists and engineers, constantly adapting to the requirements of a similarly evolving vehicle fleet. Because of the hazardous nature of the products the system requires safety features from transport to final sale. Private ownership is a system that has served us well, which grew in a very organic way from its earliest years and has enabled a degree of geographical independence and personal comfort that all of us have enjoyed. Sure, they gouged us every chance they got, but they kept us on the road. Nobody's perfect.

The new landscape, however, will see an entirely different set of features. To say reflexively that the same corporate giants which today dominate the energy arena should maintain their control over the new paradigm bears challenging. For one thing, the many difficulties and risks inherent in the fossil fuel business will be almost entirely eliminated with the GREAT system. Nuclear power research and development has necessarily been under the control of governments, never in the hands of the private companies. That is as it should be, and as it will remain. Any evolutionary improvements in that area will be developed and demonstrated in government laboratories. Private utilities that operate nuclear power plants only add to the destabilization of the system by being responsible (or irresponsible) for maintenance and training of their employees. They have been, in too many cases, derelict in both those duties. What else, pray tell, do they contribute? They aren't needed to raise money to build the plants from the sale of stock; that money can be as readily raised by bond issues, which is how publicly owned utilities raise the money they need. The teams of GREAT operators would be trained under a protocol and to a degree of expertise that would be consistent worldwide rather than subject to the tender mercies of the private utilities' bean counters who likely know squat about the technical challenges involved but have plenty of ideas about how to cut corners to save money.

When it comes to boron, here too we can see that the differences are like night and day. Unlike with oil, we don't have to keep extracting it. Once you acquire it from either known deposits or, easily enough, from seawater, you don't ever need to get any more (except a gradual increase in the vast boron pool to accommodate an increase in vehicle numbers). The recycling process is very straightforward, completely different than the ever-changing formulations and tweaks of oil refining. You're doing only one thing: taking boron oxide and driving off the oxygen. Period. No need for special weather formulations, no need to mix in ethanol or anti-knock additives, none of that complexity. In your car, boron combines with oxygen to make boron oxide. On the other end, oxygen is driven off and released back into the air (or used to generate electricity) and you've got your boron again. What possible benefit can Exxon Boron contribute to that system? It's so simple and straightforward that they'd have altogether too much free time to figure out creative ways to separate you from your money.

The transport of boron fuel, and of boron oxide back to the recycling plants, can easily be accomplished using privately owned trucking companies, since neither of those items is volatile in the least. Simple market forces would keep those prices quite stable, especially in light of the fact that fuel prices for the trucks and trains that carry the boron would no longer be variable. The extraction of the boron from land or sea sources could likewise be left in private hands, for it's quite a simple process and can be done in any country with ocean water available. If our own boron extractors decided to jack up their prices, we could just buy some from our neighbors until they brought their prices back down. Boron would be a global commodity, and once the world's vehicle fleet is converted it would be pretty much idling along from then on.

There is an imponderable here, though, which should be mentioned. We want to deal with GHG emissions as quickly and decisively as possible. The vast majority of them arise from two sources: electrical generation and vehicle emissions, the latter being only slightly less than the former. The problem with cutting the emissions from electrical generation is that the lead time to build a substantial number of new clean power plants is a lot longer than the time it would take to design and begin producing boron hybrids.

In Chapter Two I briefly alluded to a new battery technology that, if it turns out to be as good as its promoters claim, could revolutionize automobile transportation. At first glance it would almost lead one to think that the whole boron (or other less desirable alternatives like hydrogen) vehicle idea might be unnecessary. But such is not the case, for a speedy conversion to all-electric vehicles would run up against a serious shortfall of generation capacity to keep them charged — less so in France than anywhere else. Those wily French! In the short term boron hybrids would end up burning a lot of boron and not using the plug-in capability much, simply because there wouldn't be enough juice to charge all the cars. As time goes on and lots more generating capacity starts to come on line, the percentage of boron used will decrease substantially (provided the price of boron vs. electricity is about on a par in terms of cost per mile). It could well get to the point where people whose driving is usually limited to short hops might only rarely end up burning boron at all.

The same electricity generation shortfall applies to rapid conversion to boron if our boron oxide recycling is dependent upon electrical supplies. The distinct possibility that boron oxide can be reduced using plasma converters, though, alters that equation, since the power to perform the recycling would be provided not by electrical power plants but by the boron oxide and garbage streams in tandem. Thus rapid deployment of boron vehicles would be enabled, but the rationale for keeping the boron oxide recycling under GREAT control, as discussed above, would be substantially less compelling. Indeed, if oil companies decide to move into the plasma converter business, one could easily imagine them taking over the boron recycling business and thus retaining their stranglehold on vehicular energy supplies.

This is exactly the sort of scenario that so many people who've discussed this energy revolution have insisted must be sought if we are to see an energy revolution come to pass. It isn't hard to understand the resigned belief that the fossil fuel companies are so politically and economically powerful that no major changes will happen unless they retain their control over fuel supplies. Plasma conversion of boron oxide, if it can be accomplished, throws the door wide open for oil companies to remain our fuel masters. It's easy to see, from their forays into hydrogen and biofuel research, that they don't much care what the fuel is, as long as they control it.

Unfortunately, the long lead time to bring substantial new generating capacity on line, coupled with our desire to convert to boron as soon as possible, puts us over a barrel (pun intended). How our society decides to deal with the implications will be another one of many political decisions. The vast majority of the public would likely hope that some sort of preemptive antitrust policies be put in place to keep oil companies from completely co-opting the boron business. If not, they would be so entrenched by the time sufficient new generation capacity comes online that displacing even some of their boron recycling capacity with the GREAT system proposed earlier — in which boron acts as the great storage battery — would be at best difficult, and at worst politically impossible.

Actually, if the plasma conversion process does work for boron recycling then that would be the logical method to use even if we had plentiful electricity, since its efficiency would seem to be far superior to other methods. As for using boron as an energy storage buffer as suggested above, we would be thrown back on older methods (e.g. pumping water back up into hydroelectric reservoirs when electricity is in excess) or new technologies. One possible example is so-called flow batteries in which electrolytes retaining charged ions are pumped in and out of giant storage tanks that funtion as battery systems.

There would be little cause for concern if we could assume that oil companies taking over boron reprocessing (and very possibly boron extraction as well) wouldn't gouge the public, but we have little historical evidence to back up such a hope. This possibility should not, however, deter a commitment to deploy boron hybrids as quickly and pervasively as possible. Since we'd be starting from scratch with these systems, there would be no reason why independents couldn't build plasma converters and bring true competition to the boron marketplace. Perhaps the free market would work just fine in this case. One would hope that government regulators would be standing by with a big stick in case Exxon Sanitation or BP Waste Management get too pushy.

In the short term these could be potential trouble spots, but in the long term the problem would regulate itself. Once enough IFRs come online and electrical generation is sufficient to charge hybrids and all-electric cars, boron producers would be forced to keep boron prices low and stable enough to compete with the low electric rates. If not, the demand for boron would plummet as drivers switch over to battery-powered driving whenever possible.

Retail boron sales could logically be left in private hands. A certain minimal amount of specialized equipment would be required depending on which form the boron car designers ultimately settled on. But because of the stability and safety of the fuel (the antithesis of hydrogen, I would remind the reader) there would be no need for special regulations limiting just who could sell it. Virtually any type of retail establishment could be a boron vendor. As with gasoline today, prices would settle into a narrow range, probably narrower than gasoline prices since no vendor would be able to claim that their formulation is better than somebody else's. Boron vendors would make a modest profit, and customers would benefit from consistently low prices for their vehicle fuel (assuming the gougers don't take control of the supply).

Of course a large part of gas prices today is actually tax, and those taxes vary considerably from country to country. They provide funds for our transportation infrastructure — a socialized system, by the way, even in the incorrigibly capitalist USA. The taxation would undoubtedly persist in much the same way that it does today, though as a percentage of the cost it would be much higher than today because the boron fuel itself would be exceedingly cheap. Perhaps we could add a couple cents in the USA to keep our freeways free, instead of selling them off to foreign companies at fire sale prices so they can collect tolls from Americans for decades.[215] Even with higher taxes, fuel prices would tumble around the world, and remain low indefinitely. For there is no shortage of boron in the world. There's more than enough to fuel billions of cars at a time. Currently there are less than a billion vehicles in use around the world. All we have to do is extract the fuel reserve to power them and we're off to the races.

Which brings up the issue of energy conservation. Getting the American government to raise the energy efficiency requirements of vehicles, the so-called CAFE standards, has been like pulling teeth. Eschewing the use of SUVs has been seen

---

[215] Daniel Schulman & James Ridgeway, "The Highwaymen," *Mother Jones* Jan 1, 2007.

by many as a mark of virtue, leading to jokes about the well-known hybrid, the Toyota Pious. Slapping a "We Support Our Troops" magnetic ribbon on an SUV even as those troops fight and die in a war over oil supplies has come to be considered by many as a symbol of rank hypocrisy, though millions of drivers still seem oblivious to the irony. Improving gas mileage is recognized as a crucial step in both fighting global warming and reducing global tensions due to oil supplies.

Boron cars will change all that. If people want to drive around in fuel-hogging monster cars, it won't matter a bit. If they're willing to pay the negligible extra cost for their fuel, it'll be up to them. Since all vehicles will produce no emissions whatsoever, and since there will be no political tensions arising from boron's supply, everybody will be able to drive around in whatever size car they want, guilt-free. Whatever floats your boat. Even the manufacturing processes of the cars themselves won't be adding more greenhouse gases to the environment from processing the steel and other materials that go into their creation, because the auto plants will be operating with electricity. Even the giant earthmovers and other equipment that mines the iron ore, and the trains that haul it, will all be boron-powered, emitting no emissions whatsoever. Steel smelters will all be electric, as they already are in some countries. We'll have completely kicked the coke habit.

As for the automakers, they will enter a golden age. The world's entire fleet of vehicles is ripe for the plucking. All of them will have to be replaced, and drivers will be clamoring for boronmobiles since they'll be able to save buckets of money on fuel. About the only vehicle market boron might have difficulty penetrating is the motorcycle market and other small engines that power lawnmowers and chain saws and the like. Miniaturizing the boron turbine and oxygen extraction system might take a while, but in the meantime motorcycles and their kin can be the perfect market for bio/garbage (B/G) fuels, which are carbon neutral. As for airliners, Sir Richard Branson's new enterprise, Virgin Fuels, seems destined to bring the advantages of B/G fuels to the airline industry. By thus filling the fuel niches for airplanes and small engines that don't look immediately accessible to boron technology (not to say that they won't get there eventually), the conversion to a carbon-neutral world energy system will be essentially complete.

In arguing the merits of GREAT, it may seem unfair to fault the fossil fuel and utility companies for remaining true to their corporate responsibility of making as much money as possible for their shareholders (let's not forget, though, that those making the corporate decisions are almost always the holders of ridiculously large amounts of their own stock). But there's no reason to show a corporation pity. They are financial entities, not people. The individuals running them love to have their companies treated as people in the eyes of the law and to play on the sympathies of politicians and the public when it suits their purposes. But corporations embody none of the more redeeming qualities of personal ethics except insofar as it accrues to their public relations efforts and, through that, their bottom line.

The market has no conscience. The proverbial invisible hand is not held out in friendship but in a fist. It's only fair that these corporations be treated as the business entities they are and not like somebody's poor old auntie who's worried about being thrown into the street. Who benefits from privatized ownership of utilities and energy companies? Besides the overpaid executives, it's the stockholders, most of whom are well-off individuals and to whom the corporations owe their best efforts at producing maximum profit. Who provides the profit? The consumers, of course. Contrast this with publicly owned utilities that provide the model for GREAT, where construction is financed by bonds. Operating expenses (and eventual redemption of those bonds) come from consumers, but with the guiding principle that prices and service are for their benefit, rather than the benefit of stockholders.

During the California "energy crisis," which we examined in the last chapter, the state's annual power cost went from $10 billion in 1999 to $30 billion in 2000, and another $30 billion in 2001. By 2002 it got back to $15 billion, still high because of long-term contracts signed under duress in 2001. Who raked in that staggering $45 billion in excess charges? The stockholders and wildly overpaid executives of private utilities and energy trading companies who themselves contribute nothing to the system, that's who. It was a shakedown of Biblical proportions. Not content with shaking down just Californians, though, the folks at Enron (the worst of the bunch but only in degree) took even their own shareholders to the cleaners. Many of their own employees lost their entire life savings as the executives encouraged them to invest everything they had in the company's stock, even as the execs themselves pulled their own millions out in the certain knowledge of the company's impending collapse.

But it's not only in America where energy giants eat their own, as we shall see momentarily as we examine the international picture. My readers will hopefully forgive whatever tendency I have to draw my examples from an American perspective, but of course it's natural because despite having resided in other countries at various times I am, indeed, an American who's lived most of my life in the United States. The ramifications of the energy revolution I propose, however, are hardly so insular. Solving several global crises in one fell swoop will require entirely new levels of international consensus and cooperation. Such a revolution will inescapably have powerful economic, social, and political impacts on every country in the world.

CHAPTER ELEVEN

Going Global

Control energy and you control the nations.
— Henry Kissinger

W HEN DEALING WITH nettlesome problems like nuclear waste, air pollution, oil wars, and climate change, a global perspective is critical. Unfortunately, much of what we hear from the most fervent advocates of alternative energy in the USA sounds hopelessly parochial, like this is the only country that has to deal with these issues. There are countless articles predicting where we (i.e. the USA) could be by 2050 if we only would get behind solar, wind, and biofuel technologies and turn them into a juggernaut. Considering that they produce but a tiny fraction of the energy used today, it would be a transformation of epic proportions to expect that the USA could be providing all its energy from renewables within 43 years, no matter how enthusiastically the country got behind them. We'll leave off discussing the feasibility of this in general, which is far from evident.

Even in the most optimistic of such scenarios — I would venture to say *especially* in the most optimistic ones — the renewables utopia rarely seems to extend beyond America's borders (lo siento, Mexico), or at best to encompass some of the countries of the European Union. But it's called global warming, not U.S. warming. There's a reason for that. It doesn't matter how fantastically successful the USA can be in the next half century while transforming our country to an alternative energy paradise if much of the world is left muddling along with old technologies. By the time we get to 2050 we'll be in impossibly dire straits if we don't effect a truly global energy revolution, and do it soon. Even if we managed to pull it off in the USA and thus demonstrate the possibility for the rest of the world (ah, the shining city on the hill syndrome, don't Americans love it?), we'd likely have squandered our one and only chance to get global warming under control. Some think we may already have reached a point of no return. Who wants to consider the far graver possibilities half a century hence?

We've got to see over the horizon, and not only envision but implement policies and technologies that are global in their applicability. We have to get beyond the lab mentality, past pretending that we can scale up pet projects to global scale unless that sort of thing is clearly possible. We examined a variety of possibilities in Chapter Two, all of which leave grave doubts about their ability to accomplish the sort of worldwide transformation we desperately seek. The GREAT plan, however, does indeed seem feasible, provided the politicians of the world can bring themselves to commit to it. The many socioeconomic, environmental, and political changes that GREAT would bring about bear examining from a global perspective, for they will vary substantially from one part of the world to another. Let's start by looking at that other elephant in the room.

China

China is the second-largest consumer of energy in the world after the United States. Unlike the USA, though, China's energy demand is growing almost exponentially. With a

booming economy desperate for energy, the country is fever-
ishly building virtually anything short of treadmills that can
crank out electricity. In 2005, China added the equivalent of
all the power plants in Norway and Sweden to its electricity generating capacity.[216] New coal power plants are coming on line about once a week, and we're talking about the
quick and dirty pulverized coal plants, not high-tech plants.
But even at this rate it's not enough; China's calling on its
neighbors to help. There are two new coal-fired plants being
planned for Mongolia, giant ones of 3.6 GWe each, which
will send nearly all their power into China. The electricity produced by those two power plants alone is almost ten
times Mongolia's current generating capacity. The pollution
that will spew from their stacks will end up shared, albeit
unequally, by both countries, since China's northeastern arm
is downwind.

China's leaders are fully aware of their coal problem. Their
cities are already bathed in acid rain and are blessed only rarely
with blue skies. The government has introduced taxes on high-
sulfur coal to encourage a switch to natural gas and renewables, and is actually siting solar cell manufacturing plants near
abandoned coal mines. Energy efficiency is also high on their
priority list, since the Chinese know full well that money spent
on such programs gives them far more bang for the buck than
spending money on yet more power plants.

Asia's greenhouse gas emissions are expected to triple over
the next 25 years,[217] and China represents a huge part of that
global problem. According to the IEA, coal may account for between 59% and 70% of China's generation capacity in 2020,[218]

---

[216] IEA, *China's Power Sector Reforms* (Paris: International Energy Agency,
2006).

[217] BBC, "Asia's Greenhouse Gas 'to Treble'," in *BBC News International
Edition* (Dec 14, 2006).

[218] IEA, *China's Power Sector Reforms*.

with only the barest of high-tech improvements. Relatively simple sulfur scrubbers, an urgent need considering their serious acid rain problem, have only been installed on a few dozen power plants. Even under the IEA's most optimistic projections, China's already formidable GHG emissions promise to more than double. As of early 2008 it has finally outpaced the USA in carbon emissions, a leadership position of dubious distinction that the US has held since the late 19th century. China now holds the ignominious position of World's Worst Polluter.[219] It is almost beyond comprehension to think of what a doubling of that pollution will mean.

The globalization of pollution eliminates the option of dismissing this as somebody else's problem. Satellites have tracked vast clouds of pollution and dust crossing the Pacific from China and spreading over the western USA. High-altitude collectors located in the Cascade Mountains of Washington state are coated with black soot, sulfur compounds and other pollutants from Chinese coal-fired power plants.[220] The fact that the Chinese leadership is finally acknowledging their pollution problem should not in any way encourage a sanguine optimism, for they show no inclination to scrap plans for continued deployment of coal-fired power plants still on the drawing board.

The pace of development in China is perhaps difficult to appreciate for those in the industrialized countries of the west, for it has vaulted into the modern era at a dizzying speed. Just 25 years ago, fully 90% of household energy in China derived from the direct burning of coal for cooking and/or heating. By the end of the 20th century that percentage had fallen to around

---

[219] BBC, "China 'World's Worst Polluter'," (BBC One-Minute World News, Apr 14, 2008).

[220] Editorial, "We Cannot Afford This Extra Pollution," *The Daily Astorian* Jun 13, 2006.

40% and remains on the decline.[221] Offshore oil exploration has shown tantalizing hints of vast untapped reserves, while China's prodigious coal reserves remain a relatively easy, albeit shortsighted, fix for their energy demands.

It would be altogether too facile to suggest that the logical way out of the coal dilemma in China would be for the country to simply embrace clean coal technologies (CCTs). For the United States to even suggest it would be a classic case of "Do as I say, not as I do." American utility companies are rushing to build dirty coal plants as fast as they can gather the permits, in order to avoid the added investment to make their power plants cleaner under the anticipated stringent regulations. IGCC coal plants (which were discussed in Chapter Two), while promising considerable environmental advantages, haven't even managed to gain a commercial foothold in developed countries because of the costs, technical challenges, and perceived risks.[222]

Even though the Chinese know their dependence on low-tech coal technologies is hurting both their own country and the health of the planet at large, the path of least resistance continues to dominate their policies. At the most recent international meetings to hammer out a sequel to the Kyoto Accords, the Nairobi U.N. climate conference in November 2006, China insisted on being exempted from the GHG cutbacks being demanded of other industrialized countries.[223] With both China and the USA — the two largest producers of global emissions — effectively out of the picture, anything else the conference may have accomplished was symbolic at best.

---

[221] Fridley Brockett, Lin & Lin, "A Tale of Five Cities: The China Residential Energy Consumption Survey," in *Human and Social Dimensions of Energy Use: Understanding Markets and Demand* (Berkeley, CA: Lawrence Berkeley National Laboratory, 2003).

[222] IEA, *China's Power Sector Reforms*.

[223] Bradsher, "Emissions by China Accelerate Rapidly."

With all the new coal power plant construction of both the recent past and the near future, and considering that each coal plant has an expected life span of about 50-60 years, any honest appraisal leads to an inescapable conclusion. If the world is to solve the problem of GHG emissions, then China (and others, including the USA) will have to contemplate the replacement of perfectly serviceable and even relatively new power plants which have barely begun to be amortized. Is it realistic to presume that this could possibly happen in a world where the bottom line reigns supreme?

Ten or twenty years ago the very idea would have been unthinkable. But global warming and pollution politics is changing the way we view the world and our place in it, and a new reality is being forced on us by increasingly dire environmental repercussions. To call the situation a disaster could hardly be considered hyperbolic. If Sir Nicholas Stern's report is to be believed, failing to avert global warming will result in millions of displaced people and a permanent reduction of global productivity of as much as 20%. Soot alone is killing, by any realistic calculation, over half a million people every year.[224] There are few events commonly considered as disasters that can outweigh the gravity of our situation.

At the same time, China's leaders see energy shortages as one of the biggest potential threats to their national stability, and probably rightly so. With foreign exchange reserves of over a trillion dollars at the end of 2006, the country's plans to secure energy supplies for its continued growth extend around the globe and only exacerbate international tensions over energy supplies. Chinese officials have not been reluctant to state their intentions to hunt for new supplies of oil and natural gas.[225]

---

[224] Kirk Smith Ph.D., *Health Burden from Indoor Air Pollution in China* (Berkeley, CA: World Health Organization, 2003).

[225] BBC, "China Mulls Energy Reserves Spend," in *BBC News International Edition* (Dec 27, 2006).

With concerns over peak oil furrowing the brows of leaders around the world, this big new kid on the block brings added pressure to an already volatile situation.

Nuclear power now provides barely two percent of China's electricity, and their goals for the next decade or so are relatively modest but not inconsequential. This is due in part to the fact that up-front capital costs for nuclear power plants are high, and also to the fact that China lacks sizeable uranium reserves. These factors combined with easy access to coal would seem to make for dim prospects of China embracing a predominantly nuclear power scheme in the near future.

Yet the many advantages of fast reactors are not at all lost on the Chinese. Indeed, they have a small one under construction near Beijing that's due to achieve criticality in 2008, and plans for a 600 MW prototype with a target date of 2015. Russia has been cooperating with China on breeder technology since 2000, and they expect fast reactors to become the predominant design by mid century.[226] Meanwhile they are involved with Russia and South Africa in planning pebble bed reactors. For anyone concerned about long-lived nuclear waste this will be a cause for concern, for the waste issuing from pebble bed reactors is quite incapable of being recycled, and adds immensely to the volume of waste because of how it's mixed with graphite.

Despite their easy and relatively cheap access to coal, it is nevertheless not free for the Chinese or anyone else. Although nuclear power plants are expensive up front, once fast reactors have been built the fuel will be essentially free for hundreds of years. Considering the serious environmental costs of coal, if those costs were ever to be factored into the equation (and they will be) there is little doubt China would seriously consider inclusion in a global IFR program. Clearly they are looking in that direction. Coupled with GREAT's boron fuel concept as a

---

[226] WNA, "Nuclear Power in China," (World Nuclear Association, Apr 2008).

preemptive strike against looming vehicle emission problems as their society goes more mobile, the temptation to become a full international player in the program would likely be quite irresistible.

The technology sharing that is part and parcel of the GREAT proposal would allow China to move quickly to the forefront of IFR deployment, and not a moment too soon. As for their new coal plants, the stranded costs could be substantially reduced if fast reactors were to be built at existing coal plant sites. (This applies to all countries where coal plants have recently been built.) It would be possible at that point to patch the reactors into the existing turbines and auxiliary equipment, leaving only the coal burner itself as a lost investment. Where logistical considerations would make it unfeasible to do so, nuclear batteries could be employed in clusters to match the capacity of the already installed turbines. Toshiba and others are chomping at the bit to deploy nuclear batteries, with designs already at hand. While scrapping the coal burner wouldn't be a trifling matter, the ability to utilize the rest of the power plant would ease the pain, and the benefits to China and the rest of the world would be well worth the cost. The question of whether and how much the other industrialized nations might be willing to contribute to coax China in this direction bears consideration. Yes, China does have quite a wad of foreign currency reserves, but it's got a lot of development ahead of it. International encouragement must be brought to bear in order to rectify a terrible — and worsening — situation.

A tax of just ten U.S. cents per gallon equivalent of boron, levied by the USA, Canada, Japan, and the countries of the EU would pay for about 10 GW of new reactors each year. Directed toward the replacement of China's new coal-burning power plants while tapping into their turbines and auxiliary equipment as suggested above, that money could likely result in at least twice that capacity to achieve the goal of shutting down those highly polluting plants. With such an incentive provided by concerned nations, China could likely be persuaded to ante up from their own substantial cash reserves to help such a program along, especially considering their desperate pollution situation. Barring such encouragement and cooperation, it would be hard to envision China implementing a plan to replace their new coal burners on their own initiative. But with such a program, if China were to match the international contribution it could result in a replacement of up to 40 GW of coal-burning power plants per year. In short order China's coal problem — which is a problem for all of us — could be put to rest.

While it might seem tempting to sit back and wait while China's energy systems evolve gradually, the global impact of their continually increasing GHGs and other pollution warrants a more proactive approach. GREAT would provide both the technology and the incentive to accelerate China's development and more quickly eliminate their dependence on coal, one of the greatest impediments to control of this serious, and undeniably global, environmental dilemma.

India

As if the exclusion of both the USA and China from the global climate protocols wasn't enough, India is also exempt because of its status as a "developing nation."[227] Yet India's energy demands are prodigious and rapidly increasing as with its Chinese neighbor. Not only that, but India's population will soon exceed China's, and this in a nation barely a third China's size. Here again the graphs of projected energy use and greenhouse gas emissions show a depressingly steep and steady upward slope.

India is fourth in GHG emissions today, barely behind Russia and almost certain to move into third place in the near

---

[227] This term is in quotes because its specious use is one of my pet peeves. More on that in the following chapter.

future. But India is taking a considerably more aggressive pro-nuclear power position. Recently discovered uranium deposits are soon to be exploited, and with one small test breeder reactor already online, a 500 MW breeder reactor is under construction with another four planned to be operational by the year 2020.

Unlike China and its technology alliance with Russia and other nations, India's nuclear power program has been conduct-ed in virtual international isolation. The country's refusal to be bound by the Nuclear Nonproliferation Treaty blocked tech-nology sharing since the Seventies, but a recent agreement with the USA (yet to be ratified by the U.S. Congress) and overtures by France have opened the door for technology sharing in the future. India is determined to have at least 25% of its electric-ity provided by nuclear power by 2050 (it's at just 3% today), and breeder reactors are a major part of their strategy. With only moderate uranium reserves, India boasts some 25% of the world's known reserves of thorium, and they intend to utilize that in breeder reactors for future power needs.[228] They've been considering breeder technology almost since they achieved in-dependence, and now their looming energy demands are bring-ing their long-simmering breeder dreams to fruition.

India had been adamant in its refusal to bring its breeder program under the purview of the International Atomic Energy Agency (IAEA), and its problems with separating their civil-ian and military nuclear programs still threaten to torpedo the pact with the USA. Yet under the terms of that new agreement they have, for the first time, assented to IAEA oversight of their civilian program.

There is absolutely no doubt that both India and China, which contain over a third of the world's population between them, will require a significant degree of international encour-

---

[228] Thorium can capture neutrons in a nuclear reactor to become fissile U-233, which is actually a more effective fuel breeder than the more com-mon U-235 or PU-239.

agement, cooperation, and support — both financial and technological — if we are to have any hope of them making significant headway reining in their GHG emissions in the near future. A modest "international development tax" on boron fuel and/or electricity would go a long way toward assisting China and India in the development of their breeder programs. Such an involvement would serve to pull them into GREAT's international orbit, with the concomitant oversight and design standardization it would entail. I trust that the reader can appreciate the advantage of seeing every nation's breeder reactor program conforming to the safest possible standardized design. Unless an international system can be implemented that will be attractive enough for every country to participate, we will inevitably see a hodge-podge of designs with a worrisome variability in safety. And if air pollution problems are international in their effects, nuclear accidents are to be taken even more seriously. A refusal to recognize this looming threat because of an almost ideological refusal to contemplate the realities of nuclear power use in the world community is simply an unconscionable failure of responsibility. We (I refer here to the USA, Germany, and others in denial) cannot continue to pretend that breeder technology is going to go away. We can either become involved as responsible citizens of the world community to guide its development as safely as possible, or we can suffer the consequences.

Standing aloof from China and India as they journey into the realm of breeder reactor technology instead of engaging them actively in an attempt to bring some international order to such a crucial sphere of development would be short-sighted in the extreme. It seems clear that old perspectives of rivalry and competition — often to the point of war — must be replaced with a more productive view that embraces the well-being of humanity at large as the defining principle of international relations.

Russia

Russia's size alone positions it as a major contributor to GHG emissions, and indeed it is third in line, though only at about one fourth the level of the USA, and soon to be overtaken by India. Rich in both oil and natural gas, Russia is one nation that can be expected to view this proposed energy revolution with a jaundiced eye, since it would mean the end of some of its major sources of foreign income. Its customers, however, would probably take a different view, especially in light of the way the country has wielded its natural gas as a political weapon in recent years.

Russia's breeder reactor deployment is farther along than any other nation. They currently have one 600 MW breeder in operation, and before the breakup of the Soviet Union they had the previously mentioned 350 MW plant in Kazakhstan that was used for both electricity and desalination. Plans to build an 800 MW plant right next to their 600 MW one have foundered, however, due to lack of financing. Nevertheless, their intention to pursue the technology is clear, with three 800 MW breeders on the drawing board,[229] and as mentioned earlier they are working to help China develop the technology as well.

There are some disquieting aspects to Russia's nuclear program, though. One is the fact that they don't always build sturdy containment buildings to house their reactor cores. The Chernobyl plant, a primitive design that was built nowhere else in the world, had no containment building at all. Neither does their 600 MW breeder reactor. This is a glaring example of the sort of thing that the GREAT program would prevent, since nuclear power development would be carried on internationally according to standardized design guidelines and under strict supervision during all phases of construction and operation.

---

[229] Center, "Overview of Fast Reactors in Russia and the Former Soviet Union."

Another unfortunate development is due in large part to the budget shortfalls that have bedeviled the construction of new plants like the 800 MW breeder. Atomic lobbyists in the Russian parliament are now attempting to pave the way for nuclear industry privatization, and the Russian Federal Atomic Energy Agency (Rosatom) is engaged in talks with investors to create privately held nuclear plants.[230] In the absence of an international body like GREAT, how many more countries will resort to privatization, and how many private companies can be trusted with the awesome responsibility for every facet of the nuclear power industry? It has been worrisome enough to see the inconsistency of the USA's own Nuclear Regulatory Commission in the thirty years of its existence, and the way private utilities in the USA have cut corners on training and maintenance. A move to privatization in Russia would be a major step in the wrong direction.

Japan

The fifth position in the GHG emissions list is occupied by Japan. Though not a member of the so-called nuclear club, Japan has developed nuclear technology due to a high energy demand and few natural resources with which to provide that energy. Like China, Russia, and India, the Japanese are determined to pursue breeder technology. In 1994 they brought a relatively modest (280 MW) breeder reactor online, the Monju reactor.

After a year and a half of operation, a thermometer well in the secondary sodium loop broke, resulting in an impressive sodium fire. (Note: This was a far different design than the Argonne IFR, yet again illustrating the importance of superlative standardized design.) Nobody was hurt, no radioactivity was released, but the quasigovernmental organization that was run-

---

[230] Bellona.org, "Beloyarsk Nuclear Power Plant," (Dec 28, 1999).

ning Monju, PNC, nevertheless attempted to cover up the accident. Not only did they falsify reports, but they actually edited a videotape taken immediately after the accident and issued a gag order for their employees to ensure their silence.[231]

The inevitable revelation of the cover-up was a huge scandal, and Monju was consequently shut down for years. But as time passed, the Japanese nuclear industry sought a permit to reopen the plant. After a series of court battles, Japan's Supreme Court gave the go-ahead to reopen Monju. It will likely go online once again in 2008. Clearly, the intention is to develop breeder technology for more widespread deployment with full-size reactors. The Japanese and others have calculated that the technology has proceeded to the point where breeder reactors can be built more inexpensively than thermal reactors.[232] (Note that this much more economically favorable estimation was not considered in the very conservative cost estimates used earlier.) Even Toshiba's 10MW nuclear "battery" is only about 20% more expensive per megawatt than our projected price for full-size IFR plants.[233] Considering that fast reactors will provide free fuel once the startup cost is absorbed, Japan's future course is obvious.

The Monju accident is yet another example of the shortcomings of privatization of nuclear plants (PNC operated the plant relatively autonomously).

The [Japanese] nuclear power industry represents technological embarrassment and organizational failure due to greed, secrecy, corruption, shoddy

---

[231] Allexperts, *Monju* (2007 [cited]); available from http://experts.about.com/e/m/mo/monju.htm.

[232] US DOE Nuclear Energy Research Advisory Committee & Gen IV International Forum, "A Technology Roadmap for Generation Iv Nuclear Energy Systems," (US Dept. of Energy, Dec 2002).

[233] WNA, "Fast Neutron Reactors," in *Information Papers* (London: World Nuclear Association, Feb 2008).

technological practices, bureaucratic incompetence and an emphasis on production at the expense of safety — and common sense.[234]

Poor training of workers and a corporate culture that encouraged cutting corners to save money have been some of the criticisms leveled at the Japanese nuclear industry. Once again we have a situation that would be greatly improved by highly trained international independent operators. And the design of the plant might well have been improved had technology been more freely shared. GREAT would be able to learn from past mistakes of breeder operations all over the world. Japan has actually paid Russia a billion dollars for the technical documentation of their BN-600 breeder reactor. It's evident that they are eager to cooperate on fast reactor technology. Russia, Japan, France and Great Britain are sharing information on breeders now, the sort of cooperation that GREAT would institutionalize worldwide. The USA is, alas, conspicuous by its absence in that list.

Unfortunately the Japanese government has decided to take what seems to be a decidedly inferior path when it comes to the design of their reactors. Instead of settling upon small modular pool-type sodium reactors with metal fuel — reactors that both Toshiba and Hitachi are anxious to build — they've decided on large loop oxide-fueled reactors like Monju. The fact that Monju had already been built at great expense seems to have played a role in that decision. If they had chosen the small modular path with metal fuel it would have been virtually impossible to justify the reopening of Monju, for the choice would have been an admission of its inferior design. Now that the deliberative process has been carried out (at least ostensibly) and the decision made, it would be extremely difficult for Hitachi or Toshiba to

[234] Feature, "Japan's Safety Response," *Nuclear Engineering International* Dec 20, 2006.

break ranks and push for their arguably superior reactors to be built instead.

Yet just how objective was that debate in the first place? It's not a stretch to imagine that the result was a foregone conclusion. Once again this illustrates the folly of a lack of international cooperation and decision-making when it comes to nuclear power. In every country there will be political pressures that will come to bear upon the choice of technologies, just as there are in Japan and the USA. But nuclear power systems are far too consequential to be left to the vagaries of political winds. Safety, proliferation resistance, waste disposal — none of these can be considered purely national issues. When it comes to nuclear power, no nation is an island, including Japan.

With 55 nuclear reactors in operation, the Japanese are clearly looking to a future where breeder reactors will play the dominant role in establishing their energy independence. Their participation in GREAT would be about as close as you can get to a foregone conclusion, and it could help steer them onto a considerably better path, working together with the rest of the nuclear powers to implement the best systems possible, regardless of local politics. In all fairness to Japan, though, it should be pointed out that nuclear politics in the USA is still firmly stuck in the LWR era. Americans aren't even *talking* about breeder reactors.

Germany

Like Japan, Germany is a non-nuclear club nation that nevertheless has the expertise and industrial capabilities to pursue nuclear fuel reprocessing and breeder reactors if they wish. The antinuclear politics of Germany are overwhelming, however, so much so that the nation has declared its intention to abandon nuclear power altogether, even as their French neighbor takes the polar opposite view (and, ironically, sells them electricity in the bargain). Being the sixth

largest source of GHG emissions, the Germans are determined to pursue alternative energy sources to get their emissions under control. This will also entail eliminating their
coal industry, into which they've been pouring subsidies of
billions of Euros per year.

Germany and Belgium decided in the early Seventies to
build a relatively small breeder reactor near Kalkar, Germany.
It was built in fits and starts over a period of 13 years marked
by protests and government commissions demanding redesigns
and other costly delays. The entire history of that reactor is a
nightmare of politics, and it never did go online. It was finally
dismantled, adding even more costs to a project that never got
off the ground.

Germany looks to be the testing ground for a country committed to completely revamping its power systems to renewable
energy sources. While Denmark is often lauded for taking the
lead in wind power, the flip side of that is that the Danes are
heavily dependent on coal-burning power plants, with nearly
three-quarters of their electricity coming from coal.[235] Contrary
to their wholesome green reputation, Denmark's per capita carbon dioxide emissions are only about average among developed
countries.[236] Germany is really where the renewable action is.
The eyes of the world will be on them to see just how successful
they can be.

Great Britain

The British, number seven on the list of GHG emitters, also
dabbled in breeders until politics intervened. Their Prototype Fast Reactor (PFR) operated from the Seventies until

---

[235] Worldwatch Institute, "Phasing out Coal: Environmental Concerns,
Subsidy Cuts Fuel Global Decline," (Aug 25, 1999).

[236] UN Development Program, "Denmark: The Human Development Index - Going Beyond Income," (UNDP Human Development Reports,
Nov 27, 2007).

1994, but was shut down when Britain withdrew support from nuclear development. Here again it was a question of politics, not technological failings. As mentioned above, the UK government is currently cooperating with Japan, France, and Russia to pursue breeder reactor technology. It's only a matter of time.

* * * * * * *

These six nations, plus the United States, are responsible for some 60% of the world's human-caused greenhouse gas emissions. While not all are members of the "nuclear club," all of them have the technology to pursue both nuclear fuel reprocessing and breeder reactor technology, and with the sole exception of Germany the six mentioned above are either already involved or in the planning stages.

America's hope of acting as a model for other nations by eschewing breeder reactors and fuel reprocessing has clearly been unsuccessful, despite the best of intentions when that course was originally charted. Rather than encouraging stability, it now only exacerbates a fragmented international situation where several countries pursue their own nuclear programs without the benefit of technology sharing and international oversight. The American position is based not on realism but on a dangerously outmoded political calculation for domestic purposes. When it comes to either nuclear weapons or civilian nuclear power, engaging with the world is a far better course than self-deception.

Breeder reactor technology was recognized in the Fifties as being essential to fulfilling the promise of virtually unlimited power from nuclear fission. Until now the price of uranium has been so low that the path of least resistance encouraged thermal reactors and once-through fuel cycles, despite the fact that long-lived nuclear waste has been piling up around the world without any reasonable plans for its safe disposal. Now, with global warming seeming to put nuclear power back on the table despite its opponents' disapproval, fast reactor technology not only looks good from a fuel supply perspective, but it also promises to clean up the waste issue that is the legacy of the thermal reactor age.

Whereas fast reactors can handily address the fuel availability issue and the dilemma of nuclear waste, safety and proliferation resistance are clearly concerns to be addressed. As we can see from the brief sampling above, go-it-alone engineering and privatization have created problems that could easily be solved by the GREAT system. Full and open technology sharing is clearly in everyone's interest. There are over 300 reactor-years of cumulative experience with fast reactors that can be drawn upon to design, build, and operate the safest reactors possible. With GREAT's highly trained operators and inspectors, the factors that have led to dereliction and penny-pinching shortcuts will be eliminated.

The almost unbelievable power latent in nuclear fuel, while its most attractive feature, is also what makes it imperative that greed and negligence be taken out of the picture. Privatized nuclear power should be outlawed worldwide, with complete international control of not only the entire fuel cycle but also the engineering, construction, and operation of all nuclear power plants. Only in this way will safety and proliferation issues be satisfactorily dealt with. Anything short of that opens up a Pandora's box of inevitable problems.

The GREAT approach will be anathema to many, while others will readily embrace it as the logical choice for nuclear security. In country after country, private ownership of nuclear facilities has led to neglect and mismanagement. Those who hope to profit from privatization of nuclear power must be rebuffed. There should be no thought of compromise. With GREAT in control, nuclear power will be maintained at the highest level of security and at the lowest possible cost for the billions of people who will avail themselves of its benefits.

As we've seen, the seven nations that top the emissions list are all countries that already have ready access to reprocessing and breeder technologies. Because of this, there need be no proliferation concerns about unlimited deployment of IFR technology in any of them. Indeed, as we've seen earlier, IFRs will actually reduce the proliferation threat. Even if the rest of the world had no fast reactors built between now and 2050, the conversion of these seven economies to fast reactors and boron cars would eliminate over 60% of greenhouse gas emissions. If you were to project the emissions reductions from all the other nations that would have ready access to boron cars, likely 80% of emissions would be eliminated even with all the other nations' electrical and other energy systems remaining unchanged.

Yet with GREAT there is no need to deny the many benefits of IFR technology to the rest of the nations of the world, or to settle for even that 20% of the emissions they might produce. The energy embassy concept proposed earlier, with power plants specifically designed to thwart terrorism or any forced entry, would allow us all to sleep soundly at night. GREAT would have so many benefits and so few disadvantages that it would be hard to conceive of any nation refusing to join simply over the issue of the international ownership of the energy embassies. What country, after all, would look askance at unlimited and inexpensive energy supplies, guaranteed by international agreement? The inevitable movement to define energy as a basic human right will be an irresistible stimulus for a new level of cooperation and support between nations.

Fast reactor technology is not going away. It is a fact of our future. That being the case, should we be content to watch it proceed fitfully, haltingly, with profit-hungry owners steering it whichever way they see fit? Clearly the standardization of design and the oversight of GREAT are far preferable, no matter whose vested interests are being rejected. It's very simple on the face of it, when one's eyes aren't blinded by greed. The good of the many outweighs the good of the few, especially the greedy few.

GREAT can begin immediately by promoting energy efficiency programs and technology sharing worldwide, since it's been clearly shown that every dollar spent on energy efficiency programs saves two to three dollars in power plant construction costs. California has already shown the way. Just going as far as California already has (and it could go much farther) would save the construction costs and environmental insult of dozens, if not hundreds of power plants worldwide.

When it comes to building the IFRs—or any nuclear power plant—the up-front costs are admittedly high. The temptation to resort to cheaper alternatives, however, must be seen from the long view. With IFRs you just have to get over the investment hump to break into a new day of virtually free fuel. There is plenty, as we've seen, for hundreds of years even on an earth where all the primary power is derived from fission (plus whatever welcome additions can be provided by clean renewables like wind, solar, and hydropower). As we saw in Chapter 7, the actual cost over the next few decades will be even lower than a business-as-usual approach, and certainly much lower than the costs—both human and economic—of runaway global warming.

While the vast majority of countries in the world will benefit tremendously while GREAT acts as a promoter of international stability and development, it must be recognized that there are a few places where it would have the opposite effect, most notably in the Middle East. Demographic trends portend serious problems for many Middle Eastern nations even without the abandonment of fossil fuels. Their population growth rates have been so high and education so often inadequate that many of the oil-rich nations find themselves with vast proportions of marginally educated and barely employable young people.

What will happen when that demographic time bomb finds its billions in foreign income drying up? Who can doubt that many of the wealthy plutocrats who've benefited royally (literally and/or figuratively) from their oil wealth will only too happily emigrate with their billions once their source of income disappears? The flight of capital, loss of export earnings, and economic bleakness for millions of young people can hardly be viewed with anything but foreboding.

On the other hand, the Middle East is hardly a bastion of political tranquility now, nor has it been for many decades. It's likely that there will be little sympathy internationally from those who've been at the mercy of OPEC since the Seventies. But whatever the impact on Middle Eastern politics, it must be compared to the tremendous benefits to the planet and everyone on it which would come about with a wholesale embrace of GREAT. One can hardly countenance rejecting it in order to prevent the sociopolitical pain of a few nations that have been only too happy to profit from others' dependency when the shoe was on the other foot.

Indeed, one cannot help but wonder to what extent the oil giants of the Middle East themselves have willingly contributed to political instability in their own backyards in order to benefit from the escalating price of oil which is always the corollary of the violence. As the redoubtable Greg Palast has observed, "… oil companies and oil states don't make their loot by finding oil but by finding trouble." Discussing the Israeli war in Lebanon in the summer of 2006, Palast perspicaciously wrote:

> Israel's Prime Minister Ehud Olmert's approval rating in June was down to a Bush-level of 35%. But today, Olmert's poll numbers among Israeli voters have more than doubled to 78% as he does his bloody John Wayne "cleanin' out the varmints" routine. But let's not forget: Olmert can't pee-pee without George Bush's approval.

Bush can stop Olmert tomorrow. He hasn't.

Hezbollah, a political party rejected overwhelmingly by Lebanese voters sickened by their support of Syrian occupation, holds a mere 14 seats out of 128 in the nation's parliament. Hezbollah was facing demands by both Lebanon's non-Shia majority and the United Nations to lay down arms. Now, few Lebanese would suggest taking away their rockets. But let's not forget: Without Iran, Hezbollah is just a fundamentalist street gang. Iran's President Mahmoud Ahmadinejad can stop Hezbollah's rockets tomorrow. He hasn't.

Hamas, just days before it kidnapped and killed Israeli soldiers, was facing certain political defeat at the hands of the Palestinian majority ready to accept the existence of Israel as proposed in a manifesto for peace talks penned by influential Palestinian prisoners. Now the Hamas rocket brigade is back in charge. But let's not forget: Hamas is broke and a joke without the loot and authority of Saudi Arabia. King Abdullah can stop these guys tomorrow. He hasn't.

...America, Iran and Saudi Arabia share one thing in common: they are run by oil regimes. The higher the price of crude, the higher the profits...[237]

Whether you find such arguments credible or not, there can be no denying that resource wars have been fought for decades and stand a very good chance of continuing and even escalating in the future, especially with rapid growth of both economies and populations. Resource wars aren't always about oil, of course. The Israeli "incursion" into Lebanon in the summer of 2006 not surprisingly saw Israel continuing its northward push to the banks of the Litani

---

[237] Greg Palast, "Blood in Beirut: $75.05 a Barrel," in *The Guardian* (2006).

River, which they've coveted for its valuable water supply and have been fighting over for decades. (Their 1978 attack into Lebanon was called, probably not coincidentally, *Operation Litani*.)

While the threat of global warming is clearly reason enough to implement the agenda proposed in this book, the prevention of wars over resources would be a boon to mankind. Ever since the Japanese tried to gain control over their oil supply by conquering the Dutch East Indies in a desperate move that was instrumental in precipitating their war with the United States in 1941,[238] the predominant cause of resource wars has been oil. Yet with the planet's burgeoning population, and with drought conditions likely worsening due to the inescapable effects of global warming, much speculation has centered on future wars over water.

Over a billion people around the world lack safe water today, resulting in the premature deaths of millions every year. Yet this deplorable situation is dwarfed by the prediction that, if we continue with business as usual, about two-thirds of the world's population will face shortages of clean freshwater by 2025.[239] Here, too, GREAT would be able to head off almost certain strife. Along with completely defusing the oil issue, every nation would have the advantage of abundant energy supplies, with as much as necessary channeled into desalination and/or canal projects. Indeed, the Soviet breeder/desalination project in what is now Kazakhstan has already proven the concept. In fact, four different countries have an accumulated 247 reactor-years of experience with nuclear desalination.[240]

---

[238] Patrick H. Donovan, "Oil Logistics: In the Pacific War," *Air Force Journal of Logistics* (Spring 2004).

[239] IAEA, "Nuclear Desalination," in *Global Development of Advanced Nuclear Power Plants* (Vienna: International Atomic Energy Agency, Sep 2006).

[240] Ibid.

Desalination is an energy-intensive process, as is the pumping required in long-distance canal projects. Yet both these tactics will have to be employed as the world's human population continues to expand. Even without the population factor, the fact that many glaciers that have provided fresh water for millions of people are now melting is creating a crisis situation that is worsening every year. In areas where canal projects could move water from areas of plenty to areas in need, the energy demand has usually been the limiting factor. It is, after all, no great feat to build a canal. The Romans and many others were doing it thousands of years ago. Pumping the water uphill periodically, though, in order to extend the distance the water can travel, is the tough part. To get an idea of just how much energy that seemingly simple process entails, think of it the other way: consider how much electricity is provided by falling water in a hydroelectric dam.

Whether they are to be met by desalination or long-distance canals, water needs require massive amounts of electricity (or heat, in the case of some desalination systems). Unfortunately many of the people most in need of fresh water live in grinding poverty and can afford to pay little if anything for the water they so desperately need. If the world is going to avoid water wars and crises in the future, energy is going to be a pivotal factor. GREAT would supply it in abundance to any nation in its orbit. Wars over water — one of the most ominous resource shortages looming on the geopolitical horizon — can thus be preemptively resolved. "If we could ever competitively — at a cheap rate — get fresh water from salt water," observed President John Kennedy nearly 50 years ago, "that would be in the long-range interest of humanity, and would really dwarf any other scientific accomplishment."[241] We now have that capability. Hopefully we'll be wise enough to use it sooner rather than later.

---

[241] Technology Quarterly, "Tapping the Oceans," *The Economist* Jun 5, 2008.

Considering the high up-front costs of building IFRs, their deployment in the poorer nations of the world will require an international commitment by the wealthier nations, to be sure. In terms of getting global warming under control this issue can afford to be delayed a bit, since the nations most capable of affording IFRs are the same ones that are the biggest GHG culprits. In humanitarian terms, however, and in the certain knowledge that economic development requires affordable energy, it will be incumbent upon the haves to attend to the plight of the have-nots. This is an especially pressing ethical question when one considers that the harshest consequences of global warming are predicted to fall heaviest on the most disadvantaged populations. It is to be hoped that the prodigious sums of money that will be saved by the elimination of coal-fired power plants can at least be partially channeled into developing countries to establish their energy independence.

This need not be considered an entirely altruistic endeavor, for if one looks at historical demographic trends it is crystal clear that population growth rates are inversely proportional to a nation's standard of living. If we hope to get global population growth under control without resorting to draconian measures, it is in everyone's interest to assist in the development of all nations. One of the main objections to such seeming altruism has always been the reality of limited energy supplies. Raw materials, too, have been a concern, for developed nations consume an inordinately large share of resources to drive their consumerist economies.

Here again we can see that the utilization of new technologies promises to eliminate many shortages. When energy is abundant and common components of consumer items like plastics and building materials can easily be obtained from garbage or industrial or agricultural waste, the fear that the advancement of the poor nations will lead to the debasement of the rich nations no longer applies. This is no longer a zero-sum situation. Admittedly, there is a finite limit to the earth's natural resources. Yet few vital materials are in such short supply as to warrant their hoarding. Most of what we possess which defines a comfortable lifestyle can be obtained from the most common and abundant materials on earth.

If you're sitting inside somewhere reading this, take a look around at your environment. If you're a typical resident of an industrialized country, nearly everything you see that provides the basic material comforts you enjoy is made of fabric, plastic, glass, metal, or wood. The metals are usually some form of steel or aluminum, neither of which is in short supply on our planet (aluminum is the most abundant metallic element in the Earth's crust). Ditto for glass, since silicon comprises about 25% of the crust. Wood is an entirely renewable resource, though admittedly those resources have been poorly managed in many cases. Fabrics and plastics can be made from natural fibers or, with plasma converters, from garbage or other waste products that are in limitless supply. Even the walls of your home are built of materials that are, in every case, readily available and easily obtainable. They could easily be built with blocks made from plasma converter slag, the walls insulated with rock wool from the same source. Energy and recycling of materials have always been the main underlying limitations. But we can clearly see that, in reality, our energy resources are limitless, and with plasma converters virtually everything that we want to reuse can be recovered. We need only make the right decisions about how to utilize our resources. There are more than enough for everyone.

What we are seeing here is the first glimpse of a post-scarcity world, long the province of science fiction. This is a world where the basic comforts of life and the provision of unlimited energy are available to everyone on the planet. Virtually all the substances utilized by the inhabitants of the world's most advanced societies to provide their creature comforts are plentiful enough to extend those comforts to humanity at large. While certain materials are inarguably in short supply, one would be hard-pressed to think of a single one that could be considered essential to a comfortable lifestyle. Unlimited energy and plasma recycling won't exactly land us in a Star Trek future, but in many respects the post-scarcity era is within our grasp.

Many who have been understandably aghast at our era's misuse and waste of resources have posited that it would take the equivalent of two or three earths to provide everyone on the planet the same degree of physical comfort now enjoyed by those in the developed countries. The problem has not been a lack of resources, however, but a waste of resources. We can now envision a future where those deplorable habits — and those limitations — no longer apply.

This book is not meant to be a treatise on the morality of materialism. I would be the first to admit that the manifestations of rampant consumerism in the United States often verge on the ridiculous. Repudiation and condemnation of the runaway consumer culture has come to represent a mark of virtue by many who recognize the foolishness it represents. There has always been a somewhat uncomfortable balance of self-righteousness for such people, however, since even when one moderates one's consumption to what would be considered a reasonable level, the many advantages of living in a nation of plenty are part and parcel of one's environment.

When the material comforts of existence are seen as being limited, then consumption beyond one's needs does indeed carry an undeniable ethical weight. As Ralph Waldo Emerson put it lo those many years ago, "Superfluity is theft." Even when the energy and raw materials involved are plentiful, there remains the often conveniently ignored issue of the conditions under which goods have been produced, be they agricultural or manufactured commodities. It is disingenuous in the extreme to point to the abolition of slavery as evidence of the social evolution of mankind when millions of desperately poor people labor under conditions that can still honestly be considered as slavery. The fact that we don't have slaves in our home is hardly confirmation of our benevolence. The moral questions of economic fairness will not be settled by availing ourselves of the technologies promoted in this book, but should command our attention and concern indefinitely.

My point is not to justify exploitation of either human or material resources, but to point out that a transformation of energy and raw material technologies as proposed herein will present a radically transformed palette upon which to paint the picture of humanity's future. Our new course will remove the limitations by which finite natural resources and energy supplies have circumscribed our existence. Unlimited energy coupled with virtually complete recycling of materials and the production of consumer goods from plentiful or renewable resources will finally allow humanity to be unshackled from the zero-sum mentality. Raising the living standards of our billions of disadvantaged brethren will be seen as a positive development by even the most voracious consumer societies, rather than perceived with foreboding as somehow detrimental to their way of life.

Admittedly this will take some getting used to. The revolution will be not just technological and political, but psychological. The passion with which consumerism is pursued is frequently grotesque in its extremes, yet the revulsion it engenders may not be so strong when it can be viewed more as shallow foolishness than callous selfishness. Much of what is considered virtuous today will be seen more as simply a matter of personal preference in a world where creature comforts are no longer in limited supply. The concept of self-denial will have to be looked at anew. Rather than concentrating on husbanding limited resources, our attention can be turned to welcoming the rest of our fellow humans into a new reality where creature comforts are the universal norm. Abundant energy and wise use of basic resources are the keys. Clearly the technologies are already within our grasp.

This won't happen overnight, but it would be foolish to dally. The conversion of primary power systems to fast reactors will necessarily be a gradual process, which in the best-case scenario will take a few decades. Conversion of the vehicle industry to boron, however, is another story. It is entirely conceivable that boron-fueled vehicles could be driving on our highways within five years. Ironically the first boron recycling plants that would be a corollary of the conversion may end up operating with natural gas for their heat requirements, since the IFR program simply won't be able to be implemented as quickly as the boron system, and it's questionable whether existing electrical generation systems would be able to handle the increased demand of electrically powered boron recycling plants. This would, however, be only an interim fix, and would allow the vehicle fleets to get off to a quick start. If the plasma conversion method proves feasible, though, then garbage alone will provide all the energy we need for boron recycling.

Long before the conversion to boron is complete, the demand for oil will have dropped to the point where the USA, one of the world's thirstiest countries when it comes to oil, will be able to rely solely on North American supplies, resulting in geopolitical and economic realignments that will be a harbinger of things to come. Even though oil prices will surely plummet worldwide, and while the temporary price of boron recycling may well be higher than it will be once IFRs are able to provide all the power necessary to support the system, the price disparity will easily be great enough and the environmental benefits so overwhelming that boron vehicles will surely carry the day even in the near term.

The conversion to boron alone will be sufficient to make a serious impact on greenhouse gas emissions within a couple of decades. Unlike the case with breeder reactors, there will be no reason to demand compliance to international oversight. An early establishment of GREAT would also facilitate a rapid move to energy efficiency technologies worldwide even before the details of reactor design, international energy embassies, and other such aspects of GREAT are settled.

Safety and security issues are two major pieces of the GREAT plan. The plant design of the entire IFR complex will be optimized for safety, reliability, and resistance to any conceivable security threat. The most accident-proof design for a pool-type reactor such as those being proposed for the IFR complexes involves, as mentioned earlier, a below-grade installation for the reactor vessels. This physically eliminates the possibility of a loss of coolant accident as long as the coolant level is maintained at a high enough level to keep the reactor covered even in the event of the breach of no less than three vessels (two stainless steel, one hardened concrete). The earth itself would act as the final containment.

Building below grade would allow the entire structure to be topped by reinforced concrete and then covered with earth, providing security even against crashing airliners and all but the most formidable burrowing bombs. Clearly a terrorist threat would be all but impossible to launch successfully against such an installation, especially since the simple expedient of using blast doors as the only possibility of entrance would make it impossible to gain unauthorized access.

The no-man's land surrounding each IFR complex needn't be barren and inhospitable. As long as the area is clear enough to be monitored for incursion by cameras operated from the reactor complex (and remotely, by satellite link) the purpose would be served. Indeed, beyond a reasonable span of grassy areas adjacent to the complex one can easily imagine wooded areas or wetlands providing an additional buffer zone so that in the unlikely event that a GREAT strike team would be required to intervene, any civilian casualties would be avoided.

The fact that rivers frequently delineate borders is happily coincident with the fact that power plants require water for cooling purposes. From a security standpoint, locating GREAT's international energy embassies along border rivers or coastlines represents a strategic advantage in the unlikely event that the GREAT strike force would ever need to be employed to secure a power plant. A coastal assault or penetration from across a border would be far preferable, from a tactical standpoint, to penetrating the interior of a hostile nation. In many cases of non-club nations, it will be possible to confine all the energy embassies to such favorable sites, with the power transferred to the interior via the grid.

The possibility of a hostile force even attempting to gain control of an energy embassy is vanishingly slim in any event. Not only would it provide no nuclear material that could be safely removed, but it would be quite difficult, in a properly designed IFR plant, to even effectively sabotage it should an attacker manage, against all odds, to gain entry. If a rebel force were determined to interrupt the electricity supply, it would be far easier to attack the distribution networks located outside the energy embassy, destroying power towers to cripple the grid. The certainty of overwhelming retaliation for violating the international sovereignty of the embassies, along with their impenetrable design, would make their security a virtual certainty.

Even in nuclear club nations it would be logical to employ the same sort of security measures as in more obviously unstable countries. And the highly trained GREAT operating teams would likewise be employed within those countries. A universal system of professionalism and oversight will afford the greatest level of safety and security no matter where the plants are located. In addition to that, full participation in the system by the nuclear club nations will encourage the others to place their confidence in GREAT. Each country will bear the cost of all construction, operation, and maintenance, paid for from the low and stable energy bills of its citizens. The oversight, management, and operation will be subject to the international trust, performed by its professional teams.

Housing for the GREAT employees would logically be located in a compound near the plants that would be an integral part of the energy parks. Given the fact that these teams would be expected to relocate to different countries at random times as a condition of their employment, such a situation would make such moves as smooth as possible. Indeed, the houses themselves could logically be of a standardized design, even down to the furnishings. But let's keep the employees happy and avoid the temptation to make this a committee affair. A simple contract with Ikea will suffice.

These are details, of course, all of which can be easily dealt with. Indeed, even the technology and security are simple matters compared to the greatest stumbling block to a GREAT future: politics.

CHAPTER TWELVE

Political Quicksand

Everyone is entitled to their own opinion, but not their own facts.
— Daniel Patrick Moynihan

O N DECEMBER 20, 1951, four light bulbs began to glow in a remote building in Idaho. They were the harbingers of a new era, shining with the first electricity ever generated by nuclear power. Yet the scientists and engineers who witnessed the humble display weren't trying to prove that electricity could be derived from nuclear power, for it was only logical that the heat resulting from fission could produce electricity just like the heat from coal or gas. They had bigger ideas in mind.

Since the early Forties scientists working with the nascent science of nuclear power had speculated that a reactor could be built that would create more fuel than it consumed. In these early days uranium was presumed to be very scarce, so any thought of using it to produce electricity on a commercial scale was fraught, it seemed, with supply problems (the large deposits of uranium known today had not yet been discovered). By 1951 researchers were finally ready to test their ideas. The first experimental breeder reactor, with the unsurprising moniker of EBR-I, was finally ready for prime time.

The scientists at what was soon to be labeled Argonne West, in Idaho — an outgrowth of Argonne National Laboratories in Illinois, the nation's first national lab — proceeded methodically to their goal. By 1953 they were able to verify that their reactor was producing one atom of new fuel for each one that it burned, about the closest anyone had ever come to a perpetual motion machine.[242] Vindicated in their theory, they went on to design other reactors that could breed excess fuel, and within ten years they had built up the breeding capacity to 1.27 atoms of new fuel for every atom burned.

In 1964 the successor to EBR-I, dubbed EBR-II (yeah, I know, but these guys were using their creativity elsewhere), was constructed to build on the early work. By this time light water reactors were already being deployed around the world, and new uranium ore deposits were being discovered to take the pressure off fuel supplies. But the scientists at Argonne could see that if nuclear energy was to be viable far into the future the breeder reactor would eventually be necessary. Not only that, but clearly there were concerns being raised by the widespread use of thermal reactors. By 1984, the Argonne team had learned enough about nuclear power generation to feel confident that they could build a much better mousetrap.

The integral fast reactor (IFR) was the brainchild that would not only address the problems of safety, proliferation, nuclear waste, and fuel efficiency, but would actually clean up the problematic legacy of spent fuel that had been accumulating since the beginning of the nuclear power age. No loose ends were to be left for future technologies to tie up. The Argonne team, a formidable group of professionals led by Dr. Charles Till, conceived and set out to develop the energy source of the future. They succeeded spectacularly.

---

[242] Of course breeder reactors do require fuel, but they convert plentiful nonfissile material into fissile material when exposed to the neutrons in the reactor.

Even before the IFR project began in 1984, Dr. Till was a frequent visitor to Washington, his expertise on nuclear power er matters being applied to the international nonproliferation efforts of President Carter. Carter himself had a considerable working knowledge of nuclear power, having been selected by Admiral Hyman Rickover to participate in the nascent nuclear submarine program. Even though his political appointees included many who espoused an antinuclear stance, the president understood the subject enough to refrain from adopting an ideological antinuclear position, while hoping to use the influence of the United States to minimize proliferation risks. His executive order forbidding the reprocessing of nuclear fuel was, in retrospect, a vain attempt to set an example to the international community.

When Jimmy Carter issued that order, the IFR concept and the pyroprocessing technology at the heart of it had not yet been developed. Nuclear fuel reprocessing at the time did indeed separate out plutonium, and Carter recognized the threat that posed to international nonproliferation efforts. It's a certainty that neither Carter nor Dr. Till could foresee that executive order being cynically trotted out nearly two decades later as a rationale for scuttling a technology that was designed specifically to solve the proliferation and reprocessing problems.

By the time Bill Clinton was campaigning for the presidency in 1992, antinuclear sentiment had become so pervasive that the industry was stagnant, with no new plants being built. While Clinton evinced no particular interest in energy issues, he knew a good political weapon when he saw one, and taking an antinuclear stance was a no-brainer from a political point of view, especially for a Democrat. Indeed, in a New Hampshire debate he eagerly labeled a rival as "pro-nuclear" as if it was a patently absurd position. By the time he became ensconced in the White House, his course had been set.

The Argonne team had been tremendously successful in

their nuclear research for over forty years. By 1992, the IFR project was in its final stages and had achieved stellar results. The equipment to demonstrate the commercial feasibility of the final step — the pyroprocessing system — was already being put into place when Clinton took office. All the highly trained personnel to finish the project were eagerly preparing for the successful completion of one of the most impressive scientific efforts of the century, a completely integrated power generating system that would assure mankind a safe and virtually inexhaustible supply of energy for thousands of years.

But the political die had already been cast. Clinton had appointed Hazel O'Leary as his energy secretary, an appointment that looked perfectly fine from a gender and race perspective but not a particularly inspiring choice from any other view. O'Leary came in fully prepared to do her boss's bidding, and that bidding meant shutting down nuclear power research — the IFR being the obvious target — and logic be damned.

By 1994 the budget battles began, and the Argonne IFR project was slated for the scrap heap. In his State of the Union address, Clinton announced his administration's intended "cancellation of un-needed programs in advanced nuclear energy." With a coterie of antinuclear advisors behind him, Clinton had decided that the EBR-II reactor, which had exceeded everyone's most optimistic expectations and had run flawlessly for some thirty years, was to be shut down and dismantled. Charles Till, who could see the finish line of a project that had consumed years of work by hundreds of highly skilled and motivated professionals, was desperate to complete the project. He describes his meeting with the deputy director of the White House's Office of Science and Technology Policy:

"I went into DC to meet with the deputy head of OSTP, who had the lead in this, to plead the case for the continued operation of EBR-II, as the fine experimental facility it was, even in the absence of the IFR program. The most economical operation of any reactor by far, because of the long years of experience getting by on constrained budgets, and, centrally, the pride of the unique, and insightful, operating crew. The reactor by now could do almost any kind of irradiation, and had a marvelous array of experimental equipment to do it. Fusion experiments, fuel irradiations for the non-proliferation fuels program, and others as they came up, materials irradiations, etc. The equipment and gadgetry had been invented over the years to take about any experiment in stride. He looked at me in a friendly way — we knew each other well — and said, "No, it has to go. It's got to be shut down. It's a SYMBOL."[243]

The shortsighted wastefulness of throwing away such a valuable resource and scattering a world-class team of scientists and engineers to the winds strictly for the sake of political machination was breathtaking. But there was still a chance that Congress would vote to fund the project, and in that event it would take an unlikely presidential veto to kill the program. Clinton and O'Leary found their surrogates on the Hill, though, and what played out in the House and Senate chambers was an example of politics at its worst. It was reminiscent of Otto von Bismarck's famous observation: "Laws are like sausages, it is better not to see them being made."

Clinton's hatchet man in the Senate was none other than John Kerry, who came loaded for bear. During the debate on the Senate floor he put on an impassioned presentation to convince his fellow senators to terminate the IFR project. But since there were no good reasons to terminate the program — and

---

[243] Email from Dr. Charles Till to the author, Dec. 9, 2006

many reasons not to—other than political posturing on the part of himself and Clinton, Kerry pulled out all the tools of the desperate debater: misinformation, misdirection, appeals to authority, and cherry-picking of reports and data.

With an overbundance of fairness to Kerry (considering his behavior) it might be pointed out that politicians are automatically in a sticky position by virtue of the nature of their jobs. A majority of Kerry's constituents were (and probably still are) almost assuredly dead-set against nuclear power. When people vote for someone to represent them in Washington, the assumption is that their representative will faithfully represent their wishes on policy matters. So what is a politician to do if he finds himself in the awkward position of believing that a policy position the majority of his constituents finds abhorrent is actually a good thing? Does he vote according to his belief, or does he take the position that his constituents would have him take? In cases of high emotion and intense activism (like nuclear power, in many states), repudiating his constituents' wishes can lead to electoral defeat. "It was for your own good" isn't something voters want to hear when their leaders have been dismissive of their clear wishes.

I have no personal ax to grind against John Kerry (except for this issue). I voted for the guy! Not only that, but I agree with the majority of his votes in Congress. Nor am I about to put words in his mouth. I am prepared, however, to use his own words pulled verbatim from the Congressional Record, and then to refute the bogus arguments he used in order to rob them of any credibility they might conceivably have in future debates.

Those senators who successfully refuted Kerry's arguments and carried the day (alas, only in the Senate) are senators no longer. Kerry, on the other hand, is not only still there but has grown in stature and influence over the years. While I would like to think that in the intervening years he would have investigated the subject and might have a more nuanced and educated view of the different types of nuclear power, I have every reason to believe that this is not the case.

John Kerry and his wife recently published a book on the environment, *This Moment on Earth*. Intended as a tome to inspire and suggest solutions to environmental crises, the potential of nuclear power to play any role in the situation is glibly dismissed in three paragraphs. No, to call it glib would be an understatement: "Nuclear energy is carbon free, and it is also available. That is the case for considering it." Wow, with a strong case like that how could anyone resist rushing out to build nuclear power plants? The final verdict, reached two paragraphs later, is that economics, proliferation concerns, and waste disposal all have to be worked out before nuclear can be considered "a sound vision for the long-term future."

Given that his fellow-senators ably pointed out thirteen years ago that the IFR was designed specifically to solve those very problems, it's clear that John Kerry is still clinging to the outdated talking points that are the time-worn staples of antie arguments. The next IFR debate may well find Kerry still in the Senate and conceivably willing (though I would hope not) to reprise his role in the unsubstantiated demonization of the IFR. Presumably he'd use similar arguments, but that dog won't hunt no more. By the time this debate is revisited, every member of Congress will have had the chance to get the real story.[244]

So the reader will please forgive me if I seem to dwell overlong on the topic. What you're witnessing is a preemptive strike. If you've never observed a contentious debate in Congress, you might even find it entertaining in a way. Let's not forget that this was a pivotal point in a program that could already be well on its way to commercial deployment if not for its untimely demise in 1994. It's prudent, I think, to air this dirty laundry in the hope that future deliberations will be fact-based and ratio-

---

[244] The 535 copies of my book that it will cost me will be the best money I'll ever spend.

nal. Lest Kerry's bluster sound too plausible, however, let's first glance once more at the realities of the IFR program, which Dr. Till had been explaining year in and year out on the Hill, and about which Kerry should certainly have been aware:

- **Safety:** By the time of this debate the EBR-II reactor had completely proven the success of its passive safety design. It had been put through worst-case scenario tests that duplicated the sort of conditions that had occurred at both Three Mile Island and Chernobyl, and had safely shut itself down without incident, exactly as the physics of the design would dictate.
- **Proliferation resistance:** The IFR is designed so that whatever fuel enters the facility never undergoes any separation of elements that can be used in a nuclear weapon, maintaining a "too hot to handle" condition at all times due to its mixture of highly radioactive materials with the fuel. No plutonium or uranium ever leaves an IFR once it goes in the door.[245] It will all be burned as fuel to produce electricity.
- **Waste:** The waste products from an IFR plant are a fraction of what would be produced by a thermal reactor, and contain neither weapons-usable material nor other long-lived actinides. Unlike the spent fuel from thermal reactors, IFR waste will be radioactive for only a few hundred years instead of hundreds of thousands of years, and will be in an inert form that keeps it from entering the water table or polluting in any way, even in a worst-case scenario. There is so little of it that all of the waste from a plant's 50 or 60-year life could be kept in a room at the facility until the plant is decommissioned.

---

[245] Except for those reactors specifically intended to produce fuel for new IFRs, of course.

- **Expense:** While the cost of building an IFR plant will be in line with the cost of thermal nuclear plants (though very possibly less once many are deployed with the same design), the fuel to keep them running will be better than free. Once a plant has its initial startup fuel, it will use free depleted uranium to produce not only the fuel to keep itself going indefinitely, but also extra fuel to start up new IFR plants as needed.

Kerry's arguments against the IFR program have been echoed by many ill-informed, disingenuous or downright mendacious critics in the years since 1994. Here's a selection of his diatribes taken from the Congressional Record, in his own words, with my commentary in italics:

> "...The advanced liquid metal reactor [ALMR] is an expensive pork-barrel project that poses serious environmental and proliferation risks." *{Wrong on every count, as should be clear if you've read this far.}*
> "...Breeders convert uranium into plutonium, the material used to make nuclear weapons. By promoting a fuel cycle based on plutonium, the ALMR inevitably increases the risks of nuclear proliferation." *{In fact, all nuclear reactors convert uranium into plutonium, including thermal reactors. With the IFR (which by definition includes an ALMR along with the pyroprocessing facilities), the plutonium is never separated and poses no proliferation risk.}*
> "...The ALMR does exactly what the President has said we should not do—reprocess plutonium. {This is simply not true.} For this reason, the New York Times last September called for an end to funding for the ALMR which—and I quote—'produces electricity by converting uranium that can't be used in warheads into plutonium, which can.'" *{This too is a fallacy except in the most twisted*

technical sense. Yes, uranium is converted to plutonium in an IFR, but that plutonium is virtually impossible to separate (and would hardly be usable for weapons even if it were, due to its isotopic composition). Taking lessons in nuclear physics from a reporter at the New York Times is not usually a sound idea, assuming that truth is the goal.}

"... The nuclear power industry has indicated that its future depends upon the success of a new generation of advanced light water reactors." *{Virtually anyone who understands nuclear physics knows full well that the only way to assure virtually unlimited fuel is to eventually switch to breeder reactors. The new generation of light water reactors is seen as an interim step only.}*

"...A National Academy of Science report gave light water reactors the highest ranking for overall performance in its evaluation." *{There were no commercial reactors of any kind in the United States other than light water reactors. It's a sure bet that the NAS didn't rank breeders at all in such an evaluation. Keep in mind that the type of reactors Kerry is so enamoured of here are the kind that leave a virtually eternal legacy of nuclear waste that contains plenty of plutonium and uranium in the mix, and which the IFR would actually clean up.}*

"...The capital costs of producing plutonium fuel are necessarily higher than those of uranium fuel because of the extra costs of reprocessing. As a result, the price of uranium ore would have to increase fifteen-fold before the ALMR would be competitive with light water reactors." *{As we've seen before, the USA alone has enough depleted uranium already mined and milled to power the entire earth's energy needs for hundreds of years. It's free! We actually would love to get rid of it.}*

"...ALMRs will not be able to dispose of military plutonium in a timely fashion. It would take another 20 years for ALMRs to be commercially available. Then, they would have to recycle military plutonium through their reactor cores for 100 years to transmute the plutonium into fission products." *{I wish this were true! If GREAT were to start building IFRs at the rate suggested in this book, the limiting factor will be that there won't be enough fuel to start up that many plants even if we use all the military plutonium available, PLUS all the spent thermal reactor fuel. We'll probably have to restrain the pace of building IFRs until the first ones can create enough fuel for the newer ones, since even reprocessing all of our spent thermal reactor fuel won't provide enough to start the program at a gallop.}*

"...As a result of the proliferation and environmental concerns the ALMR raises, I have had to conclude that continuing research into its viability is far too expensive an indulgence for a nation groaning under the burden of $4 trillion of debt." *{As we'll see momentarily, finishing the project would have been less costly than terminating it just short of its goal, as Kerry was proposing.}*

"...Mr. President, last fall I joined Senator Bryan in offering an amendment to terminate the wool and mohair subsidy, which passed. The wool and mohair subsidy was simply a waste of money. The ALMR is a waste of money and dangerous. It is nuclear mohair." *{Insert John Kerry hair joke here.}*

All those quotes were from Kerry's pitch as he introduced his "Breeder Reactor Termination Act of 1994" on February 22, '94. In the four months between then and his presentation during the bill's Senate floor debate, he apparently didn't use the time to educate himself about his pet project (and seemingly still hasn't, thirteen years later). Here are a few of his tidbits from that debate on June 30:

"...The reality of the ALMR, the advanced liquid metal reactor, is that it is a waste and that it is a danger, that it is fiscally irresponsible, scientifically irresponsible, and irresponsible with respect to arms control and nuclear waste."

Here Kerry reads from a letter from Energy Secretary Hazel O'Leary supporting termination of the project: "No further testing of the Integral Fast Reactor concept is required to prove the technical feasibility of actinide recycle and burning in a fast spectrum reactor, such as the Experimental Breeder Reactor in Idaho. The basic physics and chemistry of this technology are established." *{Kerry seems to have been blind to the irony that O'Leary here demolishes much of his argument against the IFR where he'd previously contended that actinide recycling was a dubious project. Indeed, as O'Leary said, no further testing was necessary to prove the technical feasibility. The final step of the project, the one that Kerry and Clinton were trying to stop, was a demonstration of commercial-scale pyroprocessing to remove the last possible doubts about the commercial viability of the IFR. Previous pyroprocessing of the fuel had taken place on a smaller scale and had been entirely successful, having run five successive batches through the reactor with pyroprocessing between each run.}*

Here's Kerry lamenting the cost of the project, at the time in the range of $140 million: "We do not have anything to show for that incredible investment except running up against the barrier of nonproliferation efforts, an extraordinary amount of increased potential waste as we pursue a technology that not only puts more plutonium into circulation, but increases the amount of waste, the actinides that you then have to have in a repository and hold for literally thousands of years for it to be eliminated... My colleagues are going to come to the floor and say you can eliminate all of that because this technology is going to chew it all up. Wrong. Wrong." *{Those colleagues, whom Kerry was accusing of promoting the IFR as a pork barrel issue, were from Illinois and Idaho, where the Argonne Laboratories are located. They had taken the time to educate themselves about the IFR and could speak with authority about it, unlike Kerry. What on earth was he talking about here? The very arguments with which he's excoriating the IFR are the ones it was designed to solve. Someone is definitely wrong wrong here, but it's not Kerry's colleagues.}*

"...What this reactor does is create a reprocessing technique that is not dependent on the uranium, but separates and reuses plutonium." *{Maybe if he says it often enough it'll become true.}*

"...The technology here, even if successful, No. 1, is just not needed. We do not need this. And, No. 2, it is dangerous. It is dangerous for the very reasons that the President and Secretary O'Leary have set out: It threatens the nonproliferation protocol." *{Kerry referred to the IFR as dangerous repeatedly throughout his argument, despite the nonproliferation features of the IFR. Besides, nobody was proposing to build IFRs in non-nuclear club countries anyway. It's been suggested by some in the intelligence community that North Korea, which Kerry alluded to in his speech, already had a nuclear weapon by this time. Whether that's true or not, the fact is that they finally did build nuclear weapons from spent fuel rods.}*

"...It is only dangerous because of the questions that I have raised with respect to proliferation, to the breeder reactor and, I might add, to the additional waste that this new technology creates." *{I'm at a loss to figure out what manner of additional waste he keeps talking about. The IFR is meant to reduce waste by a factor of ten and completely eliminate the long-lived actinides.}*

At the time of this debate the long-simmering issues with North Korea were on everyone's mind. There was speculation that they already had nuclear weapons and this played into the fear mongering about nuclear proliferation. By this time, though, seventeen years had passed since Carter's directive against nuclear fuel reprocessing, and clearly the world community had no intention of following America's lead in banning it. Though Kerry and Clinton gave nonproliferation as their reason for wanting to terminate the program, the idea of setting a good example had clearly failed. In her rebuttal to John Kerry's presentation on June 30, 1994, Senator Carol Moseley Braun of Illinois drove the point home:

> "There will be some today who will tell you this is an issue about nuclear proliferation. I submit that anyone with common sense who has watched the proliferation policy in this country over the past 15 years knows we are not going to influence other nations from aspiring to or rejecting reprocessing.
>
> "For example, opposition by previous administrations had a minimal effect on reprocessing policies of major nuclear nations. France went ahead and did what it was going to do; England did what they were going to do; Japan did what they were going to do. Their argument is that they seek energy independence that we in the United States already enjoy. So the idea that we can whipsaw other nations by shutting down our research capacity really does not make a whole lot of sense and, frankly, borders on arrogance."[246]

It is certainly true that breeder reactors create plutonium, as well as other actinides and various fission products. So

---

[246] Congressional Record, Senate, June 30, 1994

do thermal reactors. A thermal reactor of 1 GW produces about 500 pounds of plutonium per year that is mixed with the spent fuel and can be extracted with PUREX technology already available to dozens of nations (including, as we have seen, North Korea). An IFR produces zero pounds (since any that comes out goes right back in again). If we are serious about eliminating the risk of nuclear proliferation as a by-product of nuclear power generation, then the entire nuclear fuel cycle should be under international control such as the GREAT program proposed here. Banning fuel recycling in the USA is no better than burying our head in the sand. Other nations will continue doing what they will.

Mischaracterizing the IFR as a breeder reactor is disingenuous as well. It can be designed to burn plutonium instead of breeding it, if that is the goal. Certainly the IFRs in the USA and other nuclear club countries should be designed and built as breeders simply because of the fact that we'll want to be creating new fuel to start up more IFRs. But any IFRs built in non-club countries could be built as burners rather than breeders. This eventuality is years away from having to be addressed anyway. The Department of Energy, in responding to an inquiry from Senator Dale Bumpers, said it would take some $60 million and some 3-plus years to convert a burner IFR to a breeder reactor. Could it be done? Yes. But when the simple PUREX process can be used to easily extract plutonium from the relatively ubiquitous spent fuel from thermal reactors, it stretches credibility to see how the IFR is more dangerous than our current situation. Again, international control would render the issue moot.

A point often lost in the proliferation debate is the fact that it would be considerably easier to create a small number of nuclear weapons quite clandestinely by resorting to small research reactors. There are dozens of them located in over 45 countries around the world, at least 25 of which already had the capability to use them to produce weapons-grade material nearly twenty years ago.[247] They are surely even more ubiquitous today. In point of fact, it's quite difficult to assemble a working bomb from plutonium — especially the plutonium derived from spent fuel because it contains mixed isotopes — perhaps the reason why North Korea's first nuclear test pretty much fizzled. Using highly enriched uranium from fuel intended for a research reactor would be much easier from an assembly perspective. While we certainly want to maintain control of plutonium and the mixtures in which it is found (such as thermal reactor spent fuel), it is ridiculous to pretend that pilfering material from commercial reactors would be the method of choice for would-be bomb makers.

John Kerry's testimony was rebutted by a few of his fellow senators, who quite effectively debunked most of his arguments. Whereas Kerry had alleged that numerous government bodies and science organizations were against the IFR program, his colleagues not only clarified how he'd misrepresented those studies but invited their fellow senators to look for themselves, obligingly having them at hand. Where Kerry had quoted several newspaper articles calling for the termination of the project, Senator Moseley Braun cheekily (in that collegial Senate manner, of course) offered to play dueling newspapers with Kerry, offering several articles from reputable papers extolling the virtues of the IFR project.

She was joined by Senator J.B. Johnston (D-LA), respected among his peers as perhaps the preeminent energy expert in the Senate at that time. Johnston had no dog in this fight. He was not from either of the states where the research was being carried out. But he knew his science, and he exposed the charade for all to see.

When the Senate floor vote finally was taken, Kerry's amend-

---

247 Cohen, "The Nuclear Energy Option."

ment to kill the IFR project was defeated. But when the House of Representatives had previously voted on their version of the bill, the project had not fared so well. A similar debate had occurred there, with the anti-IFR role taken by Congressman Coppersmith of Arizona, who cobbled together a coalition of environmentalists and budget cutters, using the deceptive proliferation argument once again to take advantage of the North Korean tensions at the time. The same sorts of arguments that Kerry had used were trotted out: proliferation, pollution, and economics. But this time the nays carried the day, sending Kerry's amendment, along with the rest of the energy bill, into a conference committee.

The administration's sleight-of-hand with the numbers had successfully confused the House members. Hazel O'Leary had flatly admitted under questioning by the House Energy & Power Subcommittee on March 9, and again in questioning by the Senate on March 23, that it would cost more to terminate the program than to finish it. Yet many House members who voted to kill the project realized only belatedly that they'd been duped. Congressman Myers of Indiana told the story:

> "...We had little choice considering the direction that the House gave us last week when the gentleman gave us instructions that we must terminate the IFR immediately. It was a mistake, no question about it. The thing that really bothers me is the fact that after that vote, a number of Members came up to me and said, 'We voted wrong. We thought we were saving money.' When they found out it was going to cost more the way we are going now to terminate this project and get absolutely nothing for the money, they [knew they] had made a mistake."[248]

---

[248] Congressional Record: August 10, 1994, Conference Report On H.R. 4506, Energy And Water Development Appropriations Act, 1995

The IFR project met its demise in the conference committee, undoubtedly in part because of Clinton and O'Leary's input. Kerry couldn't help but gloat over the fact, but others were clearly disheartened. Senator Craig ruefully remarked, "I strongly believe history will show this termination decision to be a wrong and short-sighted one." Senator Kempthorne of Idaho, now Secretary of the Interior, threw it back in Kerry's lap:

> "I believe that this decision is a mistake and I truly believe that our Nation will one day regret the Congress' decision to turn its back on this promising technology.
>
> "I supported the IFR Program as an important technology to help this nation deal with the problem of surplus weapons grade plutonium and spent reactor fuel. Now that the IFR Program will be terminated, I look forward to seeing how the critics of the IFR propose to deal with these problems."[249]

The turnaround in Clinton's stand was a complete about-face from just the year before. Senator Moseley Braun put it in perspective:

> "In one of the most remarkable moves of all, Madam President, Secretary O 'Leary this year awarded the general manager — and this is almost a funny story — the general manager of the ALMR/IFR program a gold medal and $10,000 for his work on this technology, and the Secretary at the time described the ALMR/IFR as having "improved safety, more efficient use of fuel, and less radioactive waste." So why would the administration award someone $10,000 and a gold medal for a program

---

[249] Congressional Record, Senate,Thursday, June 30, 1994

that they then turn around and want to kill, Madam President?"[250]

What really was motivating Clinton and Kerry to act so desperately to kill the IFR project? Was it their intent to pose as heroes to the environmentalist/antinuclear crowd, to bolster their green credentials? Proliferation, while clearly an issue, would just as clearly not have been worsened by continuation of the research. Indeed, it was correctly and adamantly argued by other members of Congress that it would reduce proliferation risks. Since Clinton and Gore campaigned on an antinuclear agenda, just the fact that the IFR was a nuclear program may well have been considered reason enough for killing it. The powerful fossil fuel companies surely recognized what might happen if a new form of unlimited, reasonably priced, and environmentally benign power were to be deployed. There's a lot of money on the table when you start talking about a technology that can put the fossil fuel industries out of business. The potential for such a revolution was clearly in the air, yet not a single politician ever broached the subject during the Congressional hearings on the IFR project.

There is no sure way of divining a person's intentions, of course. The cost, despite Kerry and Clinton's protestations, was definitely not an issue. The Japanese had offered to kick in $60 million to help finish the project,[251] which actually would have made it less expensive to finish than to terminate short of its goal, as O'Leary had ruefully admitted under questioning.[252] So she and Kerry used numbers that reflected not the cost of

---

[250] Ibid.

[251] "Congressman Harris Fawell (R-IL)," (Congressional Record: August 10, 1994).

[252] "Senator Paul Simon (D-IL)," (Congressional Record-Senate: June 30, 1994).

finishing the research, but the cost of taking the technology to commercialization, conflating this with the research cost figures. The old shell game.

The final irony of all this is almost painful. The anti-IFR crowd had expressed their outrage that continuation of the project would cost up to three billion dollars, yet those numbers were actually projected costs for taking it all the way to commercial viability by about 2008. Clinton himself had written to Kerry that, "The IFR has no foreseeable commercial value." Now, just 14 years later, we find ourselves in dire need of just such a technology, and the Japanese have developed it — without our help — to the point where they have shown it to be as economical as light water reactor technology (or cheaper). And that three billion dollars? For the cost of about ten days of our current resource war in Iraq, we could have been at the point where we'd be ready to start building IFRs this year! Instead we find ourselves probably eight to ten years away from that point in even a best-case scenario, with Congressional battles set to be refought over the same issue.

What is particularly galling is that both Clinton and Kerry have now taken up the cause of global warming as if they're a couple of white knights galloping up to save us from our peril. (Yes, I voted for both of them.) Kerry in particular is so exaggerated in his pose that, at the time of this writing, his web site contains a speech he gave in June 2006 called *Three New Bold Ideas for Energy Independence and Global Climate Change* containing these humble lines: "So we need a plan that actually does what the science tells us we have to do. That's why I will be introducing in the Senate the most far-reaching proposal in our history."

This has got to be good, huh? Sorry to disappoint after a buildup like that, but Kerry's bold ideas consist of ethanol production, carbon trading (the deadly international shell game for perpetuation of the status quo), improving CAFE standards (yawn), and energy efficient light bulbs. Oh, and "new technologies," conveniently hazy. I guess that would qualify as four or five ideas so I don't really understand the title, especially since none of the ideas are either new or bold in the least. As for being "the most far-reaching proposal in our history?" To quote Jon Stewart, "Eh...not so much."

Any politician will find it a daunting prospect to embrace the proposals in this book, since this energy revolution, like revolutions throughout history, is certainly going to go against the grain of a lot of people and organizations with vested interests. The fossil fuel and utility companies, of course, will be the most vehement foes, since GREAT would have them staring at their own epitaphs. But corporations and industries on the verge of obsolescence, as powerful as they are, will be buttressed in their opposition by ideologies and political structures that depend on the status quo for their enrichment. And it won't be just American interests that will want to reject it.

The bankers and multinational corporations that have danced with the G8,[253] the IMF, and the World Bank for lo these many years would be horrified to see this come to pass. Indeed, they have managed to make the grudging debt forgiveness of the poorest nations contingent upon privatization of those countries' water and power systems. Internationalization of energy production would be anathema to them.

But is there any other way to take advantage of the unlimited potential of safe newclear power — or indeed to make it as safe as we should insist it be — without taking it out of the hands of profit-driven corporations? With about 30 countries capable of extracting plutonium from thermal reactor spent fuel today, and undoubtedly more tomorrow, prudence would dictate one of two choices: ban nuclear power worldwide, or put the system under international control. Clearly a Hobson's

---

[253] Canada, France, Germany, Italy, Japan, Russia, the United Kingdom and the United States

choice, for the genie is out of the bottle. Nuclear power, including fast reactors that can be built as breeders or burners, is with us to stay. International control, technology sharing, and standardization of optimum designs is not only the safest course, but the one that will assure the greatest benefit to the greatest number of people while providing all the energy that we need into the distant future.

Of course it won't be popular with the plutocrats. We shouldn't care in the least, and wouldn't have to if not for the fact that they own, to one degree or another, many of the politicians who'll be making these decisions. Nowhere is this more evident than in the United States as we approach a presidential election year. Campaign spending on media alone in 2008 has been projected to cost over $4 *billion!*[254] You don't get billions of dollars from Mom & Pop putting five bucks in an envelope and sending it to their favorite candidate (though Obama's making a go of that strategy to an extent never before witnessed). Corporate money puts presidents and Congressmen in power, and even if there's no explicit quid pro quo, everybody knows which side their bread is buttered on.

It will be up to the G8 leaders to repudiate their benefactors and take the path that is in the best interests of humanity, for the corporatocracy can be expected to fight GREAT tooth and nail. As if that wasn't bad enough, knee-jerk antinuclear activists, willfully blind to the possibilities presented here, will also be putting the pressure on politicians just as they did during Clinton's tenure. If we get another administration in power that holds up a cross to fend off anyone saying the word nuclear, it will be time to invest in sunblock futures, because we probably aren't going to get a second chance to start fighting off the global warming monster another decade or so hence.

---

[254] Reuters, "New PQ Media Report: Campaign Media Spending to Grow 64.1% in 2008 to All-Time High," (Dec 6, 2007).

Yet many of the G8's corporations will play major parts in the energy revolution because their most powerful members, like the giant international construction firm Bechtel, will assuredly be involved in the construction of the thousands of IFR power plants. Cosmic justice would best be served if the notorious Halliburton were to be forcibly dismantled as a consequence of their war profiteering and fraud in the latest round of empire building, but they too will almost assuredly be queuing up to the trough in the great energy revolution once they find they can't stop it.

The money poured into the construction of thousands of new power plants will be prodigious, though as we learned in Chapter Seven it won't really be any more than would otherwise be spent to maintain the type of energy systems we use today. At the same time it will save literally trillions of dollars and millions of lives just in external costs compared to the systems the IFRs will replace. We can avoid the sort of blatant waste and fraud that Halliburton and others were able to engage in with relative impunity during the Iraq War. The fact that the IFR design will be standardized and simultaneously constructed by hundreds of engineering and construction firms from all corners of the globe will ensure that construction cost benchmarks will be realistic. No more free lunch for the bandits. Whereas the construction companies can be expected to make a reasonable profit while employing many thousands of workers, the universality of the individual projects will be a de facto insurance against overt corruption and waste. Allowing GREAT to play a part in the financial oversight of the construction projects even as they are financed by the countries in which they are built would serve to maintain transparency and keep the costs in line.

Agribusinesses in the breadbasket of the United States are another group who'll look askance at GREAT and a future of boron cars and plasma converters. Right now a select few of these mega-corporations are making a killing growing corn for subsidized ethanol to feed the nation's SUVs, and they don't want to see that easy money dry up. Ditto Monsanto and Dupont who supply them with seed, fertilizer, and pesticides. The abandonment of the corn ethanol scheme, however, will decrease the pressure on water supplies that farmers in the corn growing states of the USA are already noticing as the ethanol infrastructure is coming on line. The serious land use issues associated with biofuels today will no longer be a concern. All the biofuels we'll need can come from garbage and other waste via plasma converters.

Intransigent free market ideologues will predictably be apoplectic at the very idea of nonprofit internationalized energy, undoubtedly portraying it as the evil globalized socialism that has had them hiding under the covers for decades. Probably not last and certainly not least, the international arms industry is unlikely to take kindly to any notion like GREAT that promises to defuse conflicts and promote harmony among nations. They and the fossil fuel companies thrive on conflict. Their lifeblood is the ghoulish financial corollary of that which is spilled from Iraq to Indonesia, from Columbia to the Congo. When nearly a trillion dollars a year is spent on weaponry, you're going to see a lot more war. Supply and demand. We supply the weapons and, whenever necessary, create the demand.

This energy revolution is going to ruffle a lot of feathers. But not only do they deserve to be ruffled, they must be. Let there be no mistake: getting politicians to make the unprecedented choices to lead the world into a new age of clean, safe, abundant and inexpensive energy is going to be a lot harder than it should be. It would seem like a no-brainer, wouldn't you think? But the powerful and wealthy classes have always exerted political control, probably since the day money was first invented.

People show the greatest solidarity when faced with an external threat, and if our planet were threatened by extraterrestrial invaders we would quickly see a global unification to ward them off. Lacking that, perhaps the threat of global warming will take the place of aliens. But barring something that dramatic and incontrovertible, it will take prodigious unrelenting pressure from the public to shake our politicians from their old habits. It isn't going to be easy. You know who we need to recruit for this effort?

You.

CHAPTER THIRTEEN

Come the Revolution

Revolution is not an apple that falls when it is ripe.
You have to make it fall.
— Che Guevara

MOMENTS BEFORE I wrote these words, I was reading about the reaction to the early 2007 report from the UN's Intergovernmental Panel on Climate Change (IPCC), which was released yesterday. Jacques Chirac, the president of France, today proposed the creation of an international body to confront the problem of global warming.[255] Though such a modest proposal is more generalized and less ambitious than the creation of GREAT, nevertheless it's heartening to see that forty-five other nations immediately responded positively to Chirac's initiative.

Conspicuous by their absence from that list of approving countries were the United States, China, and India. The USA, world's most offensive polluter (a title just recently ceded to China), has had to be prodded every inch of the way when it comes to global warming. It's only been very recently that Bush has even acknowledged the problem at all, having pressured

---

[255] Associated Press, "France and 45 Other Countries Call for World Environmental Monitor," *International Herald Tribune* Feb 4, 2007.

scientists for years to omit references to global warming from their studies.

With the U.S. Congress taken over by the opposition Democratic Party, however, global warming is finally getting some of the attention it deserves. But decisive action will nevertheless be nigh impossible considering the moneyed interests who fund the campaigns of both major parties. The new UN report and other calls to action are finally awakening the somnolent American voters to the harsh reality of climate change, however. Even the overpoweringly influential fossil fuel lobbies might ultimately find their grasp on the reins of power to be slipping in the face of public alarm, with voters demanding concrete action to address what is increasingly being perceived as an existential threat to humanity.

Though global warming is hardly a threat unique to the United States, that nation's position is preeminent for two reasons: They've long been guilty of being the worst offender in terms of production of greenhouse gases, and their global standing affords them an opportunity to lead the way to a solution. Here, then, is a list of concrete recommendations, the first of which apply specifically to the government of the USA, the rest to the nations that would establish an international body to cooperate in addressing the global issues that have been the subjects of this book. Most of these suggestions apply to the comprehensive international body that hopefully will take shape along the lines of GREAT.

If the nations of the world can manage to put the betterment of humanity ahead of the greed of the few, these recommendations could quickly be enacted to form the basis of a comprehensive solution to global warming, nuclear waste disposal, nuclear proliferation, air pollution, and resource wars (including water wars of the future). Most of the points below would require ratification by individual governments, since GREAT would be establishing what are, in effect, a series of treaties. If the core group mentioned below (the G8 countries plus China, India and Australia) can agree to cooperate in these efforts, the vast majority of nations would undoubtedly participate, since it would clearly be in their own interests to do so. Here, then, is the action plan:

» 1 USA: Establish a successor to Congress' Joint Committee on Atomic Energy (JCAE). Within a year of the bombing of Hiroshima & Nagasaki, the U.S. Congress established what was arguably the most powerful committee in the history of the U.S. government.

> Congress gave the JCAE exclusive jurisdiction over "all bills, resolutions, and other matters" relating to civilian and military aspects of nuclear power, and made it the only permanent joint committee in modern times to have legislative authority. The panel coupled these legislative powers with exclusive access to the information upon which its highly secretive deliberations were based. As overseer of the Atomic Energy Commission, the joint committee was also entitled by statute to be kept "fully and currently informed" of all commission activities and vigorously exercised that statutory right, demanding information and attention from the executive branch in a fashion that arguably has no equivalent today.[256]

The formation of the JCAE recognized the uniquely critical nature of nuclear energy. The need for secrecy in those days was clear, yet Congress was unwilling to simply abrogate its responsibilities of oversight and policymaking, unlike in recent years. JCAE acted as a bridge between the two houses

---

[256] Christopher M. Davis, "9/11 Commission Recommendations: Joint Committee on Atomic Energy — a Model for Congressional Oversight?," ed. Government and Finance Division (Aug 20, 2004).

of Congress and between Congress and the president, allowing them to hammer out differences before bringing legislation to a vote.

One critical aspect of the JCAE was its independence from
the normal committee system. Membership on the JCAE had no
bearing on a Congressman's other duties or committee assign-
ments. Nor was there any limitation on how long a member could
serve. Thus the membership had very little turnover through the
years between its formation in 1946 and its eventual dissolution
in 1977. Those who participated on this unique committee (including Al Gore Senior) were able to achieve a degree of expertise
on the science of nuclear energy that would have been impossible
under normal Congressional committee procedures.

The effectiveness of the JCAE in nuclear matters is in sharp
contrast to the sort of confusion and demagoguery described
in the last chapter's recounting of the 1994 IFR debate, illustrating a structural weakness in the way Congress deals with
highly technical issues. As long as such issues aren't politically
weighted it probably matters little, for the committees assigned
to investigating them are able to call upon experts to sort out
the pros and cons. But when something as fraught with political consequence as nuclear power is on the line, the system can
fail miserably.

John Kerry's disingenuous vilification of the IFR project
was refuted in the Senate only because another senator who was
widely respected by his peers as an energy expert tore Kerry's
arguments apart (with a little help from his colleagues). Alas,
there was no such defender of the IFR in the House of Representatives, resulting in the death of the project in the conference committee. In order to refute Kerry's counterpart in the
House, one of the representatives would not only have had to be
conversant with the technology, but willing and able to mount
a spirited and convincing defense of the project. Such a person
was not to be found.

The fact is that nearly any of the 2,000 people working on the IFR project itself could easily have countered the spurious arguments that were thrown against it in the floor debates. But expert testimony is not available during floor debates, only the arguments of the representatives themselves. In cases where congressmen resort to deception, disinformation, or even outright lies to further their cause, their uninformed colleagues are left adrift to base their votes on their own ignorance of the facts about the issues in question.

The JCAE circumvented this built-in shortcoming of Congress, at least in all matters nuclear, by creating a super-committee that was able to weigh the technical as well as the political aspects of nuclear issues and which earned the respect of the entire Congress. The issues before us today are even more complex, and even more urgent, than those under the purview of the JCAE. While nuclear power and proliferation issues are definitely on the table, the nature of today's challenges would warrant the creation of a new super-committee. A logical moniker would be the Joint Committee on Energy and Climate.

Unlike the JCAE, secrecy is not an issue. In point of fact, transparency would be much preferred. The new JCEC would deal not with nuclear weaponry but with nuclear power and other energy issues relating to climate change. Its members would avail themselves of the knowledge and advice of top professionals in their respective fields, both before and during critical deliberations. This would serve to prevent the sort of atrocious legislative chicanery witnessed in the 1994 energy debate, and pave the way for decisive action on climate change and energy policy.

» 2 USA: Craft a Congressional resolution proposing a special meeting of the G8, plus China, India, and Australia (home to much of the world's proven uranium reserves), to propose the formation of the Global Rescue Energy Alliance Trust (GREAT).

These nations represent not only the major producers of greenhouse gas emissions but those with the technologies and political power to make the GREAT proposals a reality.

» 3 The GREAT conference must embrace four principles, from which all their actions would naturally flow:

- Global warming and nuclear proliferation are critical issues that demand unprecedented cooperation among nations.
- Affordable energy is a basic human right, and GREAT will strive to make it available to all people regardless of their economic condition.
- The welfare of humanity takes precedence over the welfare of corporations.
- All nuclear power plants, reprocessing, enrichment, and waste disposal facilities will be under the control of GREAT, with no private ownership allowed. All uranium mining and milling operations will be supervised by GREAT inspectors. The only exceptions to this will be the military operations of the group of nuclear club nations, whose weapons programs would be distinct from their nuclear power programs.

This conference would take global warming action to a level so far beyond the Kyoto Accords as to make them irrelevant. The very idea that people are concerned about what will happen when the toothless Kyoto guidelines expire in 2012 is sobering evidence of just how far politicians are from taking climate change seriously. There is, however, heartening evidence that the seriousness of global warming is finally penetrating all but the most obstinate corners of the corporate state. Embracing GREAT would be a huge leap, but concrete solutions must be found. Half-hearted symbolic gestures are worthless at this point. Time is running out. There are concrete steps that can be taken immediately to quickly make a serious impact on GHG emissions, and long-range plans that can lead us out of this pending calamity are urgently needed.

It must be acknowledged that the dissemination of nuclear technology throughout the world is a virtual inevitability. Even though the USA may presently deny it on their own turf, many other countries are moving ahead. As long as each country does so independently, the risks of fissile material falling into the hands of those who would use it for malevolent purposes is increased.

Despite the concerns of those who worry that membership in GREAT would represent an intolerable loss of sovereignty, the greater threat of nuclear terrorism and proliferation must be recognized as outweighing such fears. Until nuclear power is decoupled from nationalism we will never be able to control the spread of fissile material. It's not a question of sovereignty or ideology. It's a question of recognizing a danger and employing wisdom to remove it.

» 4 A worldwide commitment to open sharing of all types of energy-conserving technologies and programs would be the first and easiest step to making a real difference in the short term. Legislation mandating sweeping energy efficiency programs based on the successful California model should be considered the minimum, since California itself can still do much better (and is improving its programs constantly). It would be easy to craft energy legislation that goes even farther and could be implemented immediately. Think of it as California on steroids. (Yeah, some jokes just write themselves.)

» 5 Establish the global equivalent of CAFE standards for electrical devices. The institution of Corporate Average Fuel Economy (CAFE) standards for cars and light trucks in the USA pushed the technology of fuel economy after the oil shocks of the early 70s. In a similar way California has demanded continuous improvements in vehicle pollution control, generally acting as a bellwether for the rest of the United States.

GREAT could likewise establish adjustable standards for all manner of electrical devices. Once technology sharing becomes the norm, efficiency demands that keep pace with technological improvements would stimulate research worldwide and greatly reduce the increase in energy demand compared to the levels that are currently projected. High-draw appliances such as air conditioners and refrigerators are obvious targets. They could be required to meet certain energy efficiency levels or be banned from production. Virtually all electrical devices could be required to meet their own efficiency targets. The many current adapters—aka vampires—in use for everything from computers to telephones, draw a substantial amount of energy. Many continue to draw current even when their producing end isn't plugged in (as when a cell phone is taken away during the day). Yet the most efficient can sense when current is needed and can stop drawing current even when left plugged in.

Those of a libertarian mindset will almost surely bridle at the very suggestion of what they interpret as oppressive regulatory control. Yet the impact of such regulation will be quite invisible to consumers except in the diminished level of their electric bills. Manufacturers will be the ones most impacted, as they will be forced to keep their production methods current as technologies evolve. This may cause a slight increase in prices, yet the savings in electric bills would easily compensate. Even if it doesn't quite balance out and people end up paying a little bit more for some electrical gadgets, consider it the price of progress. After all, we're trying to save a planet here. That has got to be worth something.

» **6** Concurrent with adopting sweeping energy efficiency programs such as those just mentioned, consider how they will impact demand and take a long hard look at coal-fired power plants currently on the drawing board. Over the next 25 years some 93 GW of new coal-fired plants are planned for the USA alone, most of them still relying on old technology. GREAT should seek to convince governments to place a hold on new coal plant construction until such time as the group can evaluate the various options. The amount of energy savings realized through the prompt enactment of energy efficiency programs should easily overcome the claims of urgency that the coal industry and electrical utility companies will undoubtedly raise.

An objective look at the external costs of coal-burning power plants would convince any sane person to want the cursed things banned outright. With soot from coal burning power plants costing the USA over $165 billion per year and killing tens of thousands of citizens,[257] what possible justification can remain for allowing new dirty coal plants to be built at all? And that study doesn't even factor in other external costs of these plants such as the massive environmental damage done by coal mining, or the impact of the global warming gases that they're belching forth every day.

The overriding argument in favor of coal? We've got a lot of it and it's cheap (Pay no attention to the aforementioned study). Never mind that such an argument is simplistic, utterly foolish, and appallingly self-destructive. Every country that continues to build dirty coal plants and pump their poisons into the sky (I'm talking to you, USA and China especially) is culpable of poisoning the planet. It's got to stop. Now would be considerably better than later.

Qin Dahe, China's top meteorologist, responded to the IPCC report of February 2007 by warning that for China, as a rapidly developing nation, to completely transform its energy structure and use clean energy "would need a lot of money."[258]

---

[257] Pegg, "Coal Power Soot Kills 24,000 Americans Annually."
[258] BBC, "Climate Change 'Affecting' China," in *BBC News International Edition* (Feb 6, 2007).

It's time that both the USA and China came to grips with the real costs of their addiction to coal, at which time both nations would see that scrapping their coal industries would save them literally billions of dollars a year. More than sixty new coal-fired plants are on the drawing board in the USA. It would be outrageous to allow them to be built just to satisfy the stockholders of utility companies. The health of our planet is being undermined by the greed of a few powerful people whose decisions are quite literally killing us.

» 7 Incentivize widespread adoption of methane capture systems in agriculture. While methane lasts only four years in the atmosphere compared to over a hundred for carbon dioxide, it is nevertheless twenty times more harmful in its greenhouse effects. Agricultural operations produce a great deal of methane emissions, most of it from animal waste. In the United States, agricultural sources are responsible for about 3% of greenhouse gas emissions. Since most of this is methane, the global warming effect is all out of proportion to that seemingly small percentage. Methane digesters employed on dairy and livestock farms can capture methane and use it to produce heat, electricity, and vehicle fuel while greatly reducing groundwater contamination. These systems are already in use on many farms and have been proven to pay for themselves in very few years.

» 8 Remove the limits on how much electricity can be fed back into the grid by small energy producers. Distributed generation systems, including individual households that produce electricity in excess of their needs, can have electric meters installed that will run backwards when they have homemade power to spare. Generally the amount they receive for the power they produce is slightly less than what they pay for power they draw from the grid, which is reasonable considering the infrastructure costs involved. But in California (and possibly elsewhere, each state has its own rules) there is a cap of 4% that's been set on the total amount of homegrown electricity that the utilities will pay for. Thus a considerable investment in wind or solar panels, or in methane digester systems or micro hydro systems, might end up shortchanged for the power contributed to the grid.

This sort of problem is, of course, a symptom of private utilities and their profit motive. Remove the privatization of energy systems and there will be absolutely no need for such caps, since the entire system will be nonprofit and oriented strictly to the benefit of all the citizens on the grid. As solar and other power production technologies come into their own and prices drop, more and more individuals may well invest in home power systems. The internationalized nonprofit GREAT program will allow them to recoup their investment while the public at large benefits from the power they contribute. Virtually anyone could establish small power cooperatives as profit centers for their investors even while the grid itself remains publicly owned. It will soon become clear just how cost-effective such alternative energy systems actually are by the number and types deployed in this way. Looking at the projected costs of electricity as generated by different types of systems, I frankly doubt that anything will be as cheap as electricity from IFRs, but the door should most certainly be left wide open for any sort of renewable energy technologies that may come down the pike.

» 9 Fund a crash program to build boron-powered cars. The Bush administration has promised $1.2 billion for hydrogen research, including $119 million just last year. Boron engine development is small potatoes compared to the challenges of hydrogen. Once the boron car is developed, the government will have no reason to continue funding hydrogen research. It will be a dead issue. Put a couple of the big labs to work on the prototype turbines and oxygen extractor miniaturization project and see how fast they come up with a workable model.

By having the project developed at national laboratories, the resulting patents would be free to share with all auto manufacturers worldwide, the fastest way to wean the world off oil.

It would be reasonable and fair for the government(s) funding the research to set aside a development bonus — say about a million dollars (or Euros) — for each of the people contributing to the first commercially successful models. A little added incentive never hurt. Their development of this revolutionary engine would be worth many billions in terms of its benefit for mankind. All the money not poured into subsidies of oil, ethanol, and hydrogen would save many billions of dollars a year. The boron fuel cycle would require no subsidies at all. Even selling at ridiculously low prices, the boron infrastructure could easily pay for itself as it grows. All the speculation and hype about alternative fuels will be like a passing shower, all but forgotten when the sun comes out. Liquid fuel technologies will still have a place for small engines and portable fuel needs in areas off the grid and with no vehicles, but all those needs could easily be filled with carbon-neutral fuels produced by plasma converters.

» 10 Redirect some of the aforementioned subsidies to small-scale solar and other technologies that can be applicable to developing nations without established power grids. We cannot forget that indoor pollution from cooking with dung and wood is fatal to hundreds of thousands of people every year, and that such dire straits leads to deforestation and particulate pollution on a massive scale. Much of the Asian Brown Cloud is due to just such sources. Clean cooking technologies and small electrical systems would be a boon to development in the Third World. This is an obvious market for biogarbage fuels as well as solar. It's only a matter of will, of deciding to do it. Compared to the billions we funnel into oil subsidies alone, the cost would be a drop in the bucket.

» **11** Establish rigorous international standards for all types of nuclear power plants, with full access to international inspectors. This would be the start of the GREAT inspection and operational teams which will oversee all nuclear power plants in the future. Existing power plants found to be below standards would be either repaired immediately or shut down. No exceptions. If we're to expect people to be comfortable with nuclear power, it's imperative that they have confidence not only in the safety of the IFR plants that will be built in the future, but in the safety of the hundreds of thermal nuclear plants in use today, especially since many of them are being licensed to exceed their designed life spans.

» **12** Draft the energy embassy treaty and invite all nations to participate in the GREAT program. Make open inspection and monitoring of all nuclear facilities (including research reactors) contingent upon membership, with all the benefits that will accrue. It would be wonderful, but unlikely, that countries already in the nuclear club will agree to inspection of their weapons labs. We'll likely have to live with such secrecy for some time. All other facilities, in or out of nuclear club countries, should be under GREAT's purview.

» **13** Buy back all nuclear plants that are now owned by private utilities. Many such power plants in the USA were sold for a tiny fraction of their original cost, and should be purchased for no more than that, as much as the private utilities will undoubtedly try to jack up the price. Nationalization of nuclear power facilities will of course be met with howls of outrage from free market demagogues. Ignore them, knowing they only pretend to have your best interests at heart.

» **14** Establish the IFR working group under GREAT's jurisdiction. The first order of business should be an ongoing conference of the world's leading experts on fast reactors. India, China, Japan, Russia, and any other countries preparing to or already starting on fast reactor projects should be provided with the best possible design and construction advice that can be summoned up by the sharpest minds in the business. From now on, fast reactor development should be a planetary effort, not a national one, with completely open technology sharing.

» **15** Assemble a crack team of physicists and engineers to begin designing commercial IFR plants immediately. This is not a future technology leap. There are over 300 reactor-years of experience with fast breeders to draw on. The only element of the IFR complex that has yet to be demonstrated for commercial deployment is the pyroprocessing facility, and this is only because of the shortsightedness of the 1994 Congress. Such pyroprocessing has indeed been done on a smaller scale, and the technology itself on an industrial level is well understood. It should be pointed out that the scale of the on-site pyroprocessing facilities will be quite modest. The amounts of fuel processed at any time will be necessarily small, given considerations of criticality plus the fact that fast reactors produce truly prodigious amounts of power from almost unbelievably miniscule amounts of fuel. To keep a 2.5GW IFR running, it would only have to reprocess about a gallon of fuel per day. (It's solid, of course, not liquid. I use the gallon measurement only for purposes of illustrating how small an amount we're talking about.)

» **16** Since all the spent fuel currently scattered around the world will have to be moved and reprocessed, designate a few secure sites and begin moving it all as soon as practicable. In the USA, Yucca Mountain is probably capable of taking all the spent fuel already if we're just talking about a decade or so. Having already spent $8 billion on it, and since we won't need it for the eternal waste dump we thought we would, we might as well put it to some good use now. And by all means stop spending money on the thing! Let's save that $35 billion not yet dumped into that hole in the ground and use it to build the big reprocessing plant outside that'll convert the spent fuel into IFR fuel bundles. Anything leftover can be used to build some IFRs.

For all those worried about moving the nuclear waste, understand that we've been moving this stuff all over creation for decades. The USA has been the dumping ground for the Atoms for Peace program since before many of the people reading this were born. You just didn't know about it. There are very safe and secure methods to do this. Even in a worst-case scenario, a truckload of nuclear waste could never blow up like a nuclear bomb. It's too hot to be stolen by even the dumbest terrorist. If somehow some of it got out of its crash-proof canister it would have to be cleaned up, but the chances of such a thing happening are truly remote. There has never been an injury or death as a result of radiation from such transport even though it has been happening for decades. With thousands of people dying every year from coal pollution,[259] any hysteria over moving nuclear waste so that it can be safely disposed of once and for all would be sorely misplaced.

» **17** Put the pyroprocessing demonstration on the front burner. Given that this was all ready to go in 1994, we shouldn't have to wait ten years to build one. We needn't wait to have a fast reactor built to construct the prototype facility. The sooner we flesh out this part of the IFR program, the sooner we can begin building them and shutting down the coal plants. A sense of urgency should be cultivated in order to overcome the inertia of this type of government-funded research. With the future of fossil fuel industries on the line, you can bet that there will be

---

[259] Pegg, "Coal Power Soot Kills 24,000 Americans Annually."

powerful forces advising a go-slow approach. They must not be allowed to have their way.

Yoon Chang, a leading expert on fast reactor design at Argonne National Laboratories, has suggested that within seven years or less we could build a pilot reprocessing facility that can manufacture fuel assemblies for the IFRs from spent LWR fuel. The initial plant would be capable of reprocessing about 100 tons of spent fuel per year. After that, we'd want to look at both the amount of spent fuel we have and the amount of fuel assemblies we'll need to start up our ambitious IFR building program.

Before the IFR startup target date of 2015, we should be able to build a few very high capacity reprocessing facilities near the spent fuel depositories around the world. Taking advantage of economies of scale, this plan would minimize the cost of converting spent thermal reactor fuel into IFR startup fuel assemblies, as well as creating an optimal security situation. (For added security these plants could be built below ground. It's quite easy to build reinforced concrete structures in a hole that's been excavated with earthmoving equipment, then simply cover it back up. This would protect against both airliner strikes and even the most determined of terrorists.)

I know it will sound almost nonsensical after hearing the doomsayers moaning about nuclear waste all these years, but the real problem is that we don't have enough of it. If we're going to start building IFRs at the rate envisioned in Chapter Seven, we'll need more fuel assemblies than the spent fuel can provide. We'll either have to build them at a slower rate (thus missing our 2050 goal of eliminating all major sources of anthropogenic GHGs) or else augment the IFRs, in the early years, with the best of the third generation thermal reactors to make up the shortfall.

The problem is that the actinides needed for the IFR startup loading only comprise about 1% of spent thermal reactor fuel, and it takes about 5 tons of actinides to fire up a 1 GW

IFR. So if we could somehow reprocess all 300,000 tons of "nuclear waste" available in 2015, that would yield 3,000 tons of IFR fuel, enough to start up about 600 GW of the new reactors. The crash program proposed would build some 250 GW per year. Even if we add in old weapons-grade material from military programs we'll have less than three years' worth of IFR startup fuel at that rate of building, even if we could reprocess all of the spent thermal reactor waste very quickly.

If we site all the early generation IFRs in nuclear club countries and configure them all for maximum breeding capability, each of them will be able to create enough new fuel to fire up one more IFR of similar size in about 7 years. Thus for every plant built as a maximum breeder that means one more in seven years. If we could manage to meet our startup goals for the first seven years, after that the program would be completely self-sustaining. Of course even if we have only enough fuel for three years of startups at our one hundred plants (of 250 GW ea) per year rate, with maximum breeding we'd be able to consolidate new fuel so that by the fourth year we'd have enough from the first three to start up about sixty more. We'd be almost halfway there. The more IFRs come online, the more startup fuel will be available every year for new ones.

It would seem that the only way to meet our startup goals would be to ramp up uranium mining for a while. Embarking on a crash program of IFR building and uranium mining would surely drive up the price of uranium to hitherto unseen levels. But whereas uranium enrichment for LWRs only requires a 4% U-235 concentration, IFRs require 20%. The cost of that five-fold increase would be a deal breaker. Added to the increases in mining it would entail, and all the other cost factors, the saner choice would be to simply build as many IFRs as quickly as possible so that their breeding can begin in earnest, and make up the shortfall with the most sophisticated and safest LWRs, such as the Westinghouse AP-1000 or GE's ESBWR.

While this is not the perfect world scenario we might prefer, it is hardly a grim prospect. Just look at the major negatives of nuclear power today: safety, proliferation, cost, and waste disposal. These new LWRs are designed to be safer than any nuclear plant ever built. They employ passive safety systems similar to that developed for the IFR, and can be expected to perform perfectly well over the course of their service lives, especially considering that they would be under the construction and operational oversight of GREAT. Proliferation concerns would be addressed by that very same operational factor, and if necessary every one of the them could be built in nuclear club countries, with IFRs being built in both club and non-club nations.

The cost of building and operating these new plants, as we saw in the IEA/OECD study mentioned earlier, is the most economical of any generating system (except, probably, the IFR). So cost is not really an issue. You may remember that we were quite conservative when comparing the expense of the IFR building program to the Stern and IEA projections. The AP-1000 has a cost project of just about half the amounts we'd used in our calculations, about $1-1.2 billion per GW.[260] Given the modular design, this reactor is expected to have construction times as short as three years. Since GEH's S-PRISM design for an actual IFR will similarly be modular with greatly simplified systems, there's every reason to believe that it, too, could be expected to require similarly short build times, especially considering that the first installed power blocks can be started up before the others are even in place.

Another candidate for the transition phase between LWRs and all IFRs is the International Reactor Innovative & Secure, or IRIS. Westinghouse is heading up an international consortium to build these relatively low-power modular units that would contain the steam generators and primary coolant sys-

---

[260] World Nuclear Association, "Advanced Nuclear Power Reactors," (May 2008).

tems inside the pressure vessel. With power ranges of 100 to 335MW, they seem akin to the "nuclear batteries" discussed earlier, and would be prime candidates for nuclear retrofits to coal and natural gas power plants. Built to operate with 5% enriched fuel, they would be capable of using up to 10% enriched fuel for longer fuel cycles of up to eight years.[261] Design and certification are expected to be completed by about 2015.

That leaves the question of waste disposal. As with all thermal reactors, these transitional LWRs produce spent fuel with long-lived actinides. But rather than being a liability, that is now an asset. The more such spent fuel we have from the new reactors and the already-built thermal reactors still online, the faster we can build and start up more IFRs. And the faster we can do that, the sooner we can stop building LWRs and, from then into the indefinite future, build only IFRs. Even before 2030, under this scenario, we would have reached that point.

One advantage of augmenting the IFR building plan with third generation LWRs in the early years is that it's going to take probably seven or eight years, even under a best-case scenario, to begin our IFR building boom. But in the meantime new LWRs could start to be built. Having settled on one or two designs, it should be possible to license and build these plants much more economically and rapidly than has previously been the case with one-off designs and endless legal delays. The sooner we begin, the sooner we can shut the door on new coal plants. Take your pick. It's going to be one or the other. Gas-fired plants are looking pretty bad due to the spiking and volatility of gas supplies. Coal is the fuel that everybody's talking about using. With a commitment to passive safety LWRs in the near term and IFRs as soon as possible, we will have embarked upon a course that, frankly, we should have been on for some time.

---

[261] Ibid.

Of course building new LWRs implies the continuation of uranium mining for a portion of their operational lifetimes. As long as uranium mining exists, GREAT should be on the scene to provide oversight. Australia doesn't represent as grave a concern as some other countries like Congo, which has experienced significant pilferage of uranium for years. GREAT could also be tasked with oversight of mine safety procedures and even tailings disposal. In many respects GREAT would encompass all the tasks now performed by the IAEA, plus the operation and inspection of all thermal reactors currently online, as well as the coming deployments of IFRs.

Uranium mining will be on the decline within a couple of decades, since only thermal reactors will require their product. As LWRs reach the end of their usable lifetimes, they will all be replaced by IFRs. Well before the last of the LWRs reach that point, it will be possible to utilize extra fuel from IFR breeders to fabricate thermal reactor fuel, thus allowing the uranium industry to close down for good even as the youngest thermal plants continue to operate. Right now Australia, which boasts a large portion of the world's known uranium reserves, is locked in a political battle over whether to expand its mining. Canada, while not the leader in proven reserves, is currently the world's leading producer of uranium. So take heed, Australia. This is your time to shine. Use it or sit on it forever, because once IFRs have taken over, your uranium mines are going to be worthless.

Just as GREAT will serve to maintain the operational and technical integrity of nuclear power plants, its authority to oversee mining operations will do much to improve safety in the years leading up to the industry's eventual demise. While newer mines in many countries have greatly reduced the environmental impact of uranium mining by such measures as keeping tailings underwater during mining and subsequently burying them to prevent emissions of radon gas, some countries pay only lip service to mine safety. With some twenty new ura-

nium mines scheduled to come on line during the next couple
of decades, oversight should definitely be on GREAT's To-Do
list. Of course if the GREAT program as envisioned here is
implemented with relative haste, some of those mines would
never open, as there would be no need for their uranium.

» **18** GREAT should ban all nuclear power technologies that
create spent fuel unable to be reprocessed and burned in fast
reactors. This would include pebble bed reactors, the darling
of many pro-nuclear converts because of their perceived safety
factor. There are other nuclear technologies under development
that also produce spent fuel that's just as bad as the pebble
beds. The massive amount of spent fuel already on hand, which
is probably the most worrisome thing about nuclear power in
the public's mind, is all recyclable and usable in fast reactors.
It would be foolish in the extreme to begin producing nuclear
waste that is not, and there is absolutely no need to do so. IFRs
will provide all the energy the world needs at minimal cost and
with its safety assured by the very physics of the design.

Whenever a lot of money and effort has been poured into
development of a new technology, it's always painful to abandon
it if something better comes along. Many millions of dollars and
years of effort have been devoted to researching new systems of
nuclear power, from pebble beds to advanced high-temperature
reactors (AHTRs) and others. But a commitment to IFRs, the
benign and sensible nuclear alternative, will eliminate the need
for these technologies. The driving force behind the development of AHTRs has been the hope of hydrogen production
utilizing their high operating temperatures. But the hydrogen
economy is only another dead-end technological mighta-been.
With IFRs and boron providing primary power and vehicle fuel,
plasma converters can easily meet any hydrogen needs that remain to be filled, for they will produce considerable quantities
of hydrogen from MSW or agricultural waste.

It would be foolish in the extreme to continue R&D of technologies that have been outmoded before they got off the ground. It's not a question of hurting somebody's feelings. Nor should it be a question of hurting somebody's pocketbook. It's a question of the health of our planet. If any energy technology adds to the burden our planet and our progeny will inherit, then it deserves to be brought to an end. We want to leave this place better than we found it.

» **19** Begin a talent search for nuclear engineers. We are going to need a lot of trained professionals to oversee the construction and operation of thousands of IFRs. Because nuclear power has been so marginalized, few promising students have chosen this career path for lack of job prospects. Once GREAT commits to the IFR future, it's probable that a much greater number of students will realize the potential and choose nuclear engineering as a career, but it would help to get proactive with this and offer scholarships to begin the move into these fields. There's time, too, since the startup to the crash building program will take about as long as a high school graduate would take to earn a Ph.D. Between the construction and the many GREAT teams that will operate the IFRs worldwide, there will be a lot of career opportunities for smart kids who relish foreign travel. Perhaps the oil companies could use a bit of their prodigious bankrolls to pay for the retraining of some of their younger employees who will soon be out of a job as the fossil fuel industries bite the dust.

» **20** As soon as boron vehicles become available in quantity, slap a carbon tax on petroleum-derived gasoline. This will not only speed the conversion of the world's vehicle fleets to boron, but will make funds available for the sort of retraining just suggested above. Seeing the handwriting on the wall, OPEC will be in disarray as the nations that depend almost solely on oil for their wealth scramble to pump as much as they can before their markets dry up. Since demand will be declining and supply will be increasing, oil prices will be tumbling. Even with a carbon tax, gas prices will decline from current levels.

» **21** Restart nuclear power development research at national labs like Argonne, concentrating on small reactor designs like the nuclear battery ideas discussed earlier. Given the cost and difficulty of extending power grids over millions of square miles of developing countries, the advantages of distributed generation in transforming the energy environment of such countries can hardly be exaggerated. It is a great pity that many of the physicists and engineers who were scattered when the Argonne IFR project was peremptorily terminated chose to retire. Rebuilding that brain trust should be, well, a no-brainer. If one but looks at the incredible challenges those talented people were able to meet, it seems perfectly reasonable to suppose that a focus on small sealed reactor development could likewise result in similar success. Some of those working on the AHTR and other seemingly unneeded projects could well transition to R&D that fits into the new paradigm. Japanese companies are already eager to build nuclear batteries, and there should be every effort to work in concert with them and other researchers as we develop these new technologies. The options this sort of collaborative research would open up for the many varied types of energy needs around the world would be incalculable.

» **22** Use some of the money saved on fossil fuel and other fuel subsidies to fund research into desalination and canal projects. The world's burgeoning population is only going to add to the inevitable water crunch and pressures of future water wars. In order to nip them in the bud, we have to start planning for that future now. Cognizant of the fact that electricity will be both abundant and cheap for every nation, the deployment of massive desalination projects should be considered eminently workable. The breeder reactor/desalination project in Kazakhstan mentioned earlier in the book has already paved the way. With today's (and tomorrow's) technologies, we can do even better.

Long distance canal projects that have heretofore seemed untenable because of their energy demands will now be worthy of another look. In North America, Canada will be in the driver's seat since their ample quantities of excess water lie just north of the rapidly depleting Ogallala aquifer that sits under "America's breadbasket." If the farmers in the central United States are to be spared a return to dry land farming, Canada's fresh water would be the obvious solution. Time for Uncle Sam to make nice with his neighbor to the north.

» **23** Exert preemptive control over the new industry of plasma waste incineration. Create a body — preferably international, under the purview of GREAT — to regulate the disposal of slag from plasma burners, and to incorporate the electricity generated by the plasma burners into the local grids.

While the semi-socialized energy represented by GREAT's control over all aspects of nuclear power will undoubtedly stick in the craw of free market ideologues, the idea of establishing a regulatory body over plasma incinerators — a wonderful free-market opportunity — may seem like pouring salt on a wound. If there's one thing that rampant free-marketeering has taught us in the last few decades, though, it's that money is the bottom line for any corporation. A lack of oversight and regulation is asking for trouble. Government regulation serves the purpose of protecting the welfare of the public, in both near and long term. Corporations aspire to no such lofty goals. It's time to stop pretending they can be relied on to demonstrate social responsibility.

Anyone who would contend that plasma burners shouldn't be inspected and regulated should have to provide convincing proof that there would be no possibility of problems resulting from mismanagement or neglect. It's obvious that such is not the case. The unrestricted dumping of slag alone could be a problem, as could operation of faulty equipment that could conceivably result in pollution problems. The advantage of using the GREAT infrastructure is that international standards would apply, avoiding graft and inadequacy where regulation and the rule of law are marginal or nonexistent. Since plasma burners would be part and parcel of the energy grid, the justification for oversight by GREAT is built in. Who would buy the plasma operators' electricity when GREAT controls the grid? Their cooperation, however grudging, would be a fait accompli.

» **24** Promote a geoengineering initiative under U.N. or GREAT auspices to bring together the brainpower to tackle the problem of reversing the climate damage we've already effected.

Terraforming

The concept of terraforming sounds like it comes straight out of science fiction, which is probably where it first showed up. It basically refers to the process of transforming a planet into an earthlike environment. Usually this involves major modifications of its atmosphere (or lack thereof) and surface and subsurface features.

We have, inadvertently, been transforming our own atmosphere in quite serious ways during the past century or so as a result of both our technology and our sheer numbers. Even with the accelerated program of halting mankind's emissions of global warming gases that are recommended in this book, we will nevertheless produce much more by the time we reach the mid-century mark. Since carbon dioxide persists in the atmosphere for about a hundred years, the effects of what we've done will be plaguing our descendents for several generations. Add to this the very real possibility that we may well have already entered a very serious feedback loop in the Arctic regions with melting permafrost liberating massive amounts of methane. The diminishing sea ice in the Arctic is allowing that ocean to absorb substantially more heat than it did before. Even if we could magically cease all our GHG emissions overnight, we may still be in serious climatic trouble unless we can figure out a way to reverse what we've already done.

Taking terraforming out of the realm of science fiction may not be all that difficult, and if the truth be known it will probably prove to be a necessity. While the idea of eliminating GHG emissions and particulates from power plants and other sources sounds desirable, the fact is that it might make global warming worse for several generations. For particulates in the air have the effect of scattering sunlight, and that scattering phenomenon has somewhat mitigated the effects of global warming. Take away the particulates while the GHGs remain in the atmosphere and the full effect of global warming will be felt.

We have some clues, however, about how to rectify this sticky situation. We need only look at the short-term climate changes brought about by volcanic eruptions:

> When a volcano erupts, its ash reaches high into the atmosphere and can spread to cover the whole earth. This ash cloud blocks out some of the incoming solar radiation, leading to worldwide cooling that can last up to two years after an eruption. Also emitted by eruptions is sulfur in the form of sulfur dioxide gas. When this gas reaches the stratosphere, it turns into sulfuric acid particles, which reflect the sun's rays, further reducing the amount of radiation reaching the earth's surface. The 1815 eruption of Tambora in Indonesia blanketed the

atmosphere with ash; the following year, 1816, came to be known as the Year Without A Summer, when frost and snow were reported in June and July in both New England and Northern Europe.[262]

More recently, the eruption of Mount Pinatubo in the Philippines in 1991 resulted in a global cooling effect of about half a degree Centigrade for two years. But the aerosol sulfates that scattered the sunlight also wreaked havoc on the ozone layer, which reached its lowest levels ever recorded. As is commonly known from mankind's fling with chlorofluorocarbons (CFCs), destruction of the ozone layer would be fatal to most life on earth, since it absorbs much of the destructive ultraviolet radiation that is a component of sunlight.

Earth's atmosphere is not a homogenous zone, but rather a series of layers. The two that concern us most in terms of terraforming are the troposphere and the stratosphere. The troposphere is the layer closest to earth, within about the first ten kilometers (this changes a bit with the weather). The troposphere contains most of the atmosphere's humidity and is where our weather patterns play out. Ironically, ozone in the troposphere is considered a pollutant, and actually acts as a greenhouse gas about 25% as effective as carbon dioxide. Its persistence in the troposphere, though, is a mere 22 days.

The stratosphere is the layer above the troposphere, extending from about ten to fifty kilometers in altitude. The oft-mentioned ozone layer lies in the stratosphere, mainly in its lower reaches. Unlike the unpredictable troposphere, the stratosphere is quite stable, and airliners take advantage of that by flying in the lower levels of the stratosphere to avoid the turbulence below.

As might be inferred from its name, the stratosphere is somewhat layered, and there is relatively little mixing between

---

[262] Wikipedia, "Little Ice Age," (2008).

it and the troposphere. Thus, compounds that make their way to the stratosphere can stay up there quite a long time — one of the problems with CFCs, which continue to wreak havoc on the ozone layer even years after they've gotten there. It is also why ash and aerosol sulfates from volcanoes can continue to affect the planet's climate long after an eruption has died away.

Looking at the effect of major eruptions (and series of smaller ones) on global average temperatures raises the possibility of purposeful temperature modification by mimicking the effects of volcanoes. Of course we would prefer to avoid the untoward effects of ozone depletion, but the sunlight scattering effect of aerosol sulfates would probably be very desirable at this point in order to lower global temperatures enough to refreeze the Arctic regions and short-circuit the methane/permafrost feedback loop. The remaining polar bears would probably appreciate getting their ice back, too. Many of them have drowned as a result of having to swim for unusually long distances between ice floes.

There are undoubtedly many substances that could form aerosols and scatter sunlight as well or better than sulfates while not reacting with ozone. What we would really like, though, would be a sunlight scattering chemical that would combine with methane and/or carbon dioxide to precipitate them out of the stratosphere. Thus we could kill two birds with one stone: scatter sunlight to reduce its warming effect on the earth and remove greenhouse gases to repair the damage we've already done.

Of course even the dynamically stable stratosphere isn't completely stable, and whatever aerosols we do put up there intentionally would eventually dissipate, especially if they were purposely chosen to combine with methane or carbon dioxide and precipitate out. The aerosol sulfates from volcanoes keep their scattering going for just a couple of years. It would take a really massive amount of purposeful aerosol spraying to have any effect on our GHG problem.

Mount Pinatubo is estimated to have pumped about 10-17 million tons of aerosol sulfates into the stratosphere (depending on which figures you believe), and that resulted, as cited above, in a global temperature drop of about half a degree centigrade. If atmospheric chemists could come up with a substitute aerosol with the properties we want, we'd probably have to disperse nearly that same prodigious amount to achieve the desired effect. How, pray tell, might that be done practically and economically?

It seems almost a cosmic coincidence that the number of commercial airline flights in a year is about 18 million, reminiscent of the upper estimates of Pinatubo's aerosol tonnage. Since some of those flights aren't jetliners, however, it's reasonable to assume that the number of jets flying in the stratosphere as a matter of course lies somewhere in the 10-17 million flights/year range. Generally they fly between about 31,000 and 40,000 feet, in the lower reaches of the stratosphere.

There has already been ample experience in spraying aerosols from planes, and it would certainly be feasible to retrofit all commercial jets with emitters and tanks sufficient to dispense a thousand or two thousand pounds of aerosols as a passive activity during flight. The emitters could easily be automated so that they would be triggered when a certain altitude is reached. This would not necessarily cause long-lasting contrails, but of course there would be a lot of work to be done by atmospheric chemists and physicists to figure out the nuts and bolts of such a system.

Besides coming up with a chemical or suite of chemicals that will have the desired reactive effect with methane or carbon dioxide (and NOT with ozone), we would also have to determine how well they mix with those gases in the stratosphere and how long-lived the aerosols would be. We certainly wouldn't want aerosols with too long a lifetime, for in the event of a major eruption we'd want to reduce or eliminate the project while the effects of the volcano were at work. But given the very large number of jets that could easily and cheaply dispense the aerosols as a byproduct of their flights, there would be no reason to strive for anything too stratospherically persistent.

It will be somewhat tricky, of course, to figure out all the angles. Due to its layering, horizontal mixing of gases in the stratosphere is quite rapid, while vertical mixing between the layers is not. It's conceivable that we'd want the airliners to fly a bit higher than normal, more in the higher end of their usual range than the lower end, which often skips along the top of the troposphere.

If there's anybody I haven't managed to alarm yet in the course of this book, the idea of purposefully tinkering with the atmosphere might just send you over the edge. But given all the non-purposeful yet damaging tinkering we've already done, we may have little choice but to explore these options. We most certainly should explore them. The terraforming project would logically be approached cautiously, and once terminated would rectify itself quite readily.

******

Boldness and prudence are not necessarily mutually exclusive. In this and the other recommendations in this book, we will require an ample helping of each. Let's face it: we've messed up our planet pretty badly. It's going to take creativity and real leadership to tackle and solve our global problems. It often seems that both are sorely lacking. We'll need a leap of social and political evolution to get beyond narrow personal and nationalistic concerns. The suffocating inertia of corporatism, now unchecked even by national boundaries, will be a far greater impediment to resolving these problems than any technological challenges. But resolve them we must.

Afterword

There is no squabbling so violent as that between people who accepted an idea yesterday and those who will accept the same idea tomorrow.
— Christopher Morley

I T MAY SEEM hopelessly utopian to imagine that the idea of creating an internationalized system of energy production can be realized in the foreseeable future. Yet there are tremendously compelling reasons to create such a system. The shadowy threat of nuclear proliferation and terrorism virtually requires us to either internationalize or ban nuclear power. But a ban is impossible. At the same time, global warming is forcing us to revolutionize energy production or face the consequences. Is there a single proposed solution with any grounding in reality that excludes nuclear power? We have yet to see anything close, though we're constantly being asked to envision vaguely sketched dreams of a renewable energy future with the details conveniently omitted. Indeed, if we could only draw energy from vacuous fantasies we'd be all taken care of.

The twin threats of nuclear proliferation and global warming form an inextricable duo compelling us to consider internationalization of energy not as one option among many but as perhaps our only effective option. In terms of international cooperation, such a system would have no detrimental impact whatsoever on participating nations' sovereignty. The negligible concession to international control of energy embassies would be dwarfed by the overwhelming benefits of participation in a system like GREAT. Once that step is taken — once corporate greed is taken out of the energy equation — a whole new world is revealed in all its shining promise.

We must recognize that issues that heretofore might have been considered the problems of other nations have become our own problems. Human population has reached beyond a level where isolationism is a realistic option. Air pollution has already reached the point where millions of people cooking their dinners over choking wood or dung fires in southeast Asia ends up dumping particulates on the cities of the United States. Millions of people scrounging wood for their cooking fires have stripped huge areas bare in Africa and Asia, with resulting dust clouds spreading around the world as desertification increases every year.

Globalization is more than a question of economics. Globally shared environmental and social problems require bold international solutions. Certainly the repudiation of corporate interests in the dissolution of fossil fuel industries is, by any measure, a drastic step. These are, after all, the most powerful companies in the world. But I defy anyone to propose a solution that comes anywhere close to solving even one or two of the vast range of challenges that GREAT will enable us to solve.

Look, for example, at the crushing poverty of developing nations. Allow me to sketch just one possible outcome that could flow freely from a GREAT system where profit is no longer the motivating factor in the realm of energy.

Back in Chapter Ten I pointed out Paul O'Neill's contention that it's useless to talk about developing nations if they are going to be lacking in electricity and clean water. The clean water part of that problem is ridiculously easy and inexpensive to solve, and only lingers because of a lack of political will. O'Neill, Bush Jr.'s first Secretary of the Treasury, clearly understood and articulated how the water problem could be solved, and in fact urged the administration to initiate a pilot demonstration project. It never happened.

Electricity is a thornier problem, but with GREAT in effect the solution would be easily at hand. Recall, if you will, the relative cost of electrical generation from various types of systems. Using the most reliable figures available while being extremely conservative in our projections, we saw that IFR-generated electricity may well cost as little as 2¢ (or perhaps 2 €ents) per kilowatt-hour. This takes into account all the costs entailed in building the power plants, operating and fueling them, and ultimately decommissioning them and safely disposing of the waste products in an environmentally responsible manner.

The average electricity cost to the consumer in the USA today is about 10¢/kWh. In order to disarm critics who may quibble with my earlier 2¢/kWh production cost estimate despite the conservative foundations it was built upon, and with France's proven cost of 3 €ents to buttress our argument, let's assume that IFR electricity will cost us 4¢ per kWh. Without the profit motive built into the system, consumers should have to pay no more than about six cents even considering the infrastructure and administrative costs. Now imagine that every industrialized country with per capita income above a certain threshold would levy an international development tax (as briefly suggested earlier) of two cents per kWh on all their customers. For nearly everyone in the countries affected by the tax (Japan, USA, Canada, Australia and New Zealand, and most of Europe, among others), their electric bills would be less than they're paying now. Since the vast majority of electrical demand is in these very countries, the GREAT development fund would quickly be filled to overflowing. Construction of both power plants and electrical grids in the neediest nations would be able to proceed apace, at literally no cost whatsoever to the poor in-habitants therein.

In bringing electrification to developing countries it would probably be advantageous to utilize the expedient of nuclear batteries, which as we've seen are already knocking on the door. Certainly with international cooperation in further de-velopment there would be a wealth of options along those lines. Monolithic power plants could be built near large cities, but the nuclear battery option would allow the electrification of rural areas much more quickly and inexpensively than would other-wise be possible.

Anyone who's maintained a home and taken a trip to the hardware store to buy electrical switches, sockets, and other paraphernalia knows that the cost for such fundamental ele-ments of home electrical systems is ridiculously low. Basic elec-tric cookstoves are astoundingly cheap when considering their usefulness and longevity. A quick search online turned up a simple double-burner tabletop stove for under $20. Buy several million of them at a time and you could get a substantially better deal.

The point of this exercise, as you probably can guess, is that we want to be able to provide clean cooking facilities and basic lighting to as many poor people in the developing countries as possible, even if that means giving them away (and the electric-ity to run them) for free. This is hardly unreasonable when you consider that half the people in the world live on less than two dollars a day. If we provided those three billion people with basic cooking and lighting, that would amount to about six hundred million households, figuring family size to be three children and two parents, probably close to the norm in these situations. If a basic lighting and cooking set costs $20 per household, a quite realistic estimate, we'd be talking about $12 billion dol-lars for all three billion people. Of course we couldn't feasibly provide every one of them with electricity, some are just too far afield or in extremely sparsely populated areas, or even nomadic. But many millions — surely a large majority, in time — would be able to avail themselves of these basic amenities, at a probable cost of less than ten billion dollars for the whole program (plus electricity). Just to put that into perspective, that's about a month of Iraq war fighting, or a few weeks of just the USA's cost of dealing with coal soot. (Yes, America, burning coal costs more than fighting a war. Think about it.)

Clearly the money would really not be an issue. It might even make sense to throw in a cheap microwave with the package, since they use less electricity than electric tabletop burners. A key feature of any such program should be units that are easily repairable by the owner, with standardized spare parts (such as the microwave generator within the oven, door, hinges, basic on/off control module) in a light-gauge stainless steel body. Properly designed, they would last for many years, and be simple and cheap to repair. It would be pretty much a one-time expense. Tens of millions, even billions of people would be liberated from smoky homes and desperate daily searches for fuel. A vast amount of air pollution both inside and out would be eliminated, to everyone's benefit.

Of course it would make sense to build appliances that way in every country, but that flies in the face of consumerism and planned obsolescence. Okay, it can be argued that we can afford such nonsense. The have-nots cannot, so let's resist the temptation and try to do this sensibly at least for them. We wouldn't really have to worry about a black market on the microwaves in the developed nations, as long as we don't put a popcorn button on them. Nobody in the wealthier nations would buy a microwave without a popcorn setting, would they?

If we're going to be giving these things away, though, it would make sense to ask for something in return. How about if the recipients of this largesse return the favor by participating in tree planting projects and care of the seedlings until they mature? We could effect a wholesale reversal of deforestation in many of the most devastated areas, with all the environmental benefits that entails. Many jobs would be created in the process. People could be trained in the basics of the lineman trade to wire the villages and homes. Tree nurseries would spring up with field agents coordinating tree-planting projects throughout their countries, and any number of small businesses that rely on electricity would be made possible.

The cost of building power plants and whatever infrastructure is needed to accomplish such rural electrification would be quite easily borne by the 2¢/kWh international development tax. After all, that's really about all that building the plants and supplying the electricity would cost, and the developed countries use so much more electricity than the others that it would take decades for the developing nations to outgrow the subsidy (bearing in mind that the financially able in these countries would still be expected to pay for their power). The tax would quite painlessly maintain a never-ending source of funds to provide the poorest of the poor with not only the basic equipment but also the free electricity with which to use it, at least until such time as their level of prosperity reaches the point where they could begin bearing some of the cost without undue hardship. Both clean water systems—even if all that means is a hand pump in each village—and electricity would come to be considered human rights instead of the perquisites of the other three billion.

Contrast this with the direction that the world is going today. Powerful and heartless multinational corporations are pushing hard for electricity and water privatization in the poorest nations of the world, assuring continued and even exacerbated poverty and more needless suffering, struggle, and death for the most desperate people on the planet. How on earth have we reached this point? Even the most callous individual should feel a stirring of rage when contemplating this situation. All of us should demand that it change!

The new realities that GREAT and its spinoff programs could easily bring about can serve to not only solve our most intractable environmental problems, but to elevate the condition of the most needy among us. The zero-sum mindset that has acted as an impediment to such assistance no longer applies. There will be plenty of energy and raw materials for everyone; nobody need suffer privation. Instead of a world of haves and have-nots, we can easily and painlessly create a world of haves and have-mores. That'll do for starters. Are you okay with a beginning like that? I'm quite sure the have-nots will feel pretty good about it.

It bears repeating that such a vastly improved world will actually cost us much less, economically, environmentally and politically, than carrying on with business as usual. Nobody will suffer from taking such a course of action. The only losers will be stockholders in energy companies, who now live on substantially more than two dollars a day and will not, I assure you, have any chance of truly suffering as a result of these policies. Naturally there will be some economic dislocation as employees of fossil fuel companies (and eventually uranium companies) shift into other lines of work, but this will be a gradual process whose impact can be eased by such expedients as carbon taxes and/or boron taxes dedicated to job training in different industries. Many of the facets of oil refining today — and the jobs — will still remain, the only difference being that the source of the raw materials will be garbage and industrial or agricultural waste. It would be the height of foolishness to eschew such a course simply to keep the fossil fuel companies in business. There must come a point where common sense and compassion can win out over raw greed. Have we not come that far as a species?

The obstacle isn't technological; it's the intransigent inertia of the status quo. It's the vested interests that can't bear the thought of losing their control and their heaps and heaps of money. It's antiquated one-size-fits-all ideologies that are incompatible with a system that puts the people and the planet first. It's organizations and individuals who are willing to watch millions die in grinding poverty and in wars over oil and gas rather than abandon their profiteering. It's the billions of everyday people who live with the naïve assumption that their leaders will figure out what's best for them and take action to protect them.

Have those leaders and captains of industry figured all this out yet? Very possibly. This isn't rocket science. Your author didn't invent cold fusion in his basement. Breeder technology has been around since the Fifties. There were plenty of people even then who realized that one day it would be there to provide a virtually limitless source of energy. As for boron, people were experimenting with it as a jet fuel right after the Korean War. It could hardly be more straightforward: Boron plus oxygen equals boron oxide plus LOTS of heat. Boron oxide minus oxygen equals boron. Rinse and repeat. Granted, the idea of burning boron in pure oxygen that forms the heart of this concept wasn't conceived until recently. I wish I could say I thought of it; it was a brilliant inspiration. Oxygen extraction technologies that can make it happen haven't really had a lot of work because frankly there hasn't been much of a need for miniaturizing them. But clearly the technology is there today. Compared to the technological challenges of a hydrogen economy it's child's play. We can do this now.

Or can we? Can we overcome petty nationalism and self-defeating ideologies (I'm talking to you, Uncle Sam) and put the betterment of mankind and our planet first? Can our leaders who were elected to their positions with the collusion of their fellow plutocrats cast aside the policies that are the underpinnings of their great wealth and make decisions with a far more noble purpose? Preservation of one's lifestyle is far from noble when accompanied by disregard for one's fellow man. But embracing the preservation of our planet's health and the benefits to all mankind that it carries in its wake is noble enough, thank you very much.

Even with the very real possibility of global disaster staring them in the face, can we expect politicians to repudiate their wealthy benefactors who, to one degree or another, placed them in power? Sam Rayburn, an icon of American politics, had a bit of advice for a freshman Congressman that is pretty applicable in this situation: "Son, if you can't take their money, drink their whiskey, screw their women, and then vote against 'em, you don't deserve to be here."

It's not just because I'm an American by accident of birth that I find myself wishing that an American leader would take up this cause. Like it or not, the USA is tremendously influential on the world stage, despite its many ignoble actions of the past and present. Even with the harm done to its reputation and world standing by its recent policies, there could well be a sea change in its global standing if enlightened policymaking were embraced regardless of corporate pressures. But frankly I have little hope of that happening. The power politics in America runs way too deep. I think it's much more likely that some other nation will be the first to recognize a way out of our planetary dilemma and try to drag the USA along. I would be delighted to be wrong.

But I know I'm not wrong about the fact that it *can* be done. This book has not delved into every issue, about many of which entire books have been written. The estimates used herein have at times been necessarily imprecise, yet consistently conservative and sufficient to make the point. This is both an invitation to embrace a solution that will work and a plea to abandon costly and unnecessary dead-ends. We don't have to figure out how to economically extract oil from shale. We don't have to spend trillions on a hydrogen economy that might well prove as hazardous to our environment as the global warming we're trying to stop. We don't have to figure out how to burn coal cleanly just because there's a lot of it in the ground, and hope that pumping trillions of cubic meters of carbon dioxide into the earth won't result in a massive, deadly planetary belch. And we surely don't need to raze rainforests to grow biofuels to run our SUVs.

Leave the coal in the ground and forget about it. Plug up the oil and gas wells and turn your gaze to the future. Abandon the international jockeying for position over energy supplies, with armies ready to clash over resources that nobody even needs anymore. Embrace a new energy internationalism that will bring security and a firm base for sustainable development to all nations. There's no reason to be apoplectic (as some most certainly will be) about surrendering national sovereignty. International cooperation on critical issues has been a hallmark of political reality for centuries, and in no way demands a dissolution of nationality. Look at the creation of the European Union as a pertinent example. Such cooperation has only rarely been global in scope, but humankind's institutions are inexorably evolving and at this point that evolution is looking more and more like an absolute necessity.

The technological revolution proposed herein is entirely feasible. The revolution in consciousness is already well under way. There are billions of people who would be more than ready to embrace a spirit of international amity and cooperation promising a better life for all. Even the most hard-hearted industrialist would surely be happier to see blue sky than grey. The children of coal barons suffer from asthma too. Even if a person doesn't believe that global warming is a threat, who can wish for a world of choking pollution, international strife, crippling energy prices, grinding poverty for over half of humankind, and a legacy of toxic waste for our progeny? There's something here for everyone. So what is the great impediment?

Money. Money and power.

This revolution can be a peaceful one, indeed it must be. But revolt we must. Consider the options. Listen to the arguments. Weigh the evidence. And if you come to the conclusion that there is indeed a solution to these weighty problems, then demand that your leaders explain to you why they are not acting.

It would be wonderful to be able to provide a list of concrete actions that my readers could take to change their lifestyle in some way to contribute to the solutions this book has presented. But this isn't a matter of screwing in more twisty light bulbs and searching the want ads for a Prius. While I wouldn't discourage such actions, you certainly don't need me to suggest them. They are the province of virtually every global warming crusader out there.

What you have read in these pages constitutes an actual plan, not a feel-good exercise to nibble around the edges of global crises. This program is meant to actually solve problems, not manage them. The big issues addressed here are going to require national and international commitment by our leaders, political action of unprecedented decisiveness. If you are one of those who simply throw up their hands in resignation when faced with political causes, you can still at least discuss these ideas with others to help raise awareness of both the problems and the possible solutions.

The only way that politicians can be brought around to the point where they'll forsake the corporations on whose patronage they depend is in response to unprecedented pressure by the public. That pressure can never be brought to bear until people realize that a realistic way to solve these problems even exists. Hopefully by now it's clear that there is, indeed, just such a viable plan. In all modesty, I'm afraid I have not seen another one, and I dare say I've looked harder than you have. The closest I've seen to solutions look like energy smorgasbords: a little solar, a dash of wind, a bit of carbon sequestration, a serving of biofuels, perhaps a grudging assent to a pinch of nuclear (no breeders, please, the word is rarely mentioned and usually then only in condemnation). Will such "solutions" really get us where we want to go? A seriously qualified "hardly" is about the most honest answer. Is that good enough for you when you know that there's actually a plan that can not only work, but work quickly and solve a host of other problems in the bargain? I should hope not.

If you prefer a clear plan, if you recognize that many of the items in that smorgasbord carry their own risks and unacceptable costs, then you have to tighten the screws on your politicians. At a minimum, educate your friends and family so at least they'll understand that there is a solution. That in itself is a big step, for until now there's been a decided hopelessness in the face of seemingly insurmountable problems. If you're not a total political cynic (and who could blame you if you are?), I implore you to do your utmost to confront the politicians and demand that they take a stand. Tell them straight out that you'll vote for the ones who value real solutions over wishful thinking.

There are so many experts, so many pundits, so many politicians, so many scientists. Don't be intimidated by the letters after their names. Don't accept their stalling and equivocation. They will blather on from now till doomsday with platitudes and theories and assurances. Have you heard politicians applauded for "speaking out against global warming" as if it's a political movement or an item in the budget? Utter nonsense!

The United States is a leader in stalling and tossing up dubious proposals that even then they only tepidly support. We have the International Partnership for a Hydrogen Economy. Unnecessary and fiendishly difficult and expensive, definitely not ready for prime time. How about the Carbon Sequestration Leadership Forum? Sorry, we don't need that either. How about the Methane to Markets initiative? Since most of the methane they're talking about comes from fossil fuel production, eliminating fossil fuels will eliminate the bulk of the problem and allow us to focus that program much more effectively on other methane sources like landfills and agriculture. And yes, we've got the landfill problem covered.

In light of the plan outlined in this book, try this exercise: Read some articles on energy and climate issues. Whether it's a report on the latest global warming conference or an explanation of technologies we're being asked to consider, you'll find that most of the ideas being bandied about are so hazy and generalized as to be virtually worthless, or if there are any even remotely specific they become irrelevant or even laughable when compared with the planetary prescription presented here.

The more complex we make it, the more we're stalled into inaction. Speak out. Demand that your leaders act in the best interests of humanity, not those of their financial backers. Don't believe people who tell you it can't be done. It can. And it must.

Embrace the revolution.

Acknowledgements

T HERE'S NOTHING LIKE writing a book to give a person an appreciation for the sincerity of what's found on an acknowledgements page. Of course a reader expects to see the obligatory wife and kids references or the like, probably looking at that as perfunctory for domestic tranquility. In reality, though, mates and/or kids may well have engaged in countless discussions, listened to brainstorm musings even when they weren't interested in the least, or been disinterested observers of countless conversations rehashing ideas that they've heard repeated, through years of permutations, until they can practically write the book themselves.

A book like this goes through quite a gestation period, so when I express my appreciation to my wife and kids it's as much of an apology as a sign of appreciation for their patience. But it's more than that, for in the course of a thousand conversations large and small, their ideas imperceptibly — and sometimes quite perceptibly — contributed to the evolution of the book.

It's been about eight years since the ideas that form the core of this work first occurred to me. The first real scientist to offer me feedback on the concept was Dr. Richard Mattas, now retired but formerly the manager of the Fusion Power Program at Argonne National Laboratory. His enthusiasm encouraged me to pursue and develop them until they finally ended up on these pages.

Richard was but the first of several Argonne-auts who guided me with infinite patience to the product which you hold in your hands. Paul Pugmire, formerly the Director of Public Affairs at Argonne Labs West (since rechristened Idaho National Laboratory), fielded countless questions as I sought to bring the picture into focus, hampered as he was by a DOE directive to refrain from actually publicizing the technology whose details I relentlessly pried from him. In the intervening years the details of these technologies have been laid bare by the Internet's revelatory nature, but at the time it sometimes felt like pulling teeth. Thanks, Paul, for putting up with me and providing the answers for my blizzards of questions.

More recently, three extraordinary individuals, all participants in the Integral Fast Reactor (IFR) research at Argonne Labs, have given generously of their time and knowledge to educate me and assure the accuracy of those substantial portions of this book that deal with nuclear technology, and with the IFR project in particular. Dr. George Stanford was the first to come to my aid, and gave tirelessly of his time and applied meticulous attention to detail to assure the accuracy of my work, as well as offering welcome suggestions to file off some of the rough spots. When my questions began to lead into the intersection of science and politics, George introduced me to Dr. Charles Till, who directed the IFR project for a decade and was the voice of Argonne in the many Congressional hearings over the course of its regrettably truncated lifetime.

Finally, when discussion of some of the more esoteric points of the IFR got a bit ambiguous, Charles and George pulled Dr. Yoon Chang into our brainstorming sessions. I was struck by how both of these able scientists seemed all too ready to defer to Yoon when debatable points arose, until Charles informed me that Yoon was pretty much the go-to guy for anybody in the world who wanted to know about this technology. Like his compatriots, Yoon was a font of information and as patient as a saint. It has been both a pleasure and an honor to work with these three exceptional gentlemen, all of whom have my utmost respect and undying gratitude.

Another key to the energy revolution was provided by Graham Cowan, who somehow managed to envision an alternative fuel vehicle concept that is as inspired as it is surprising. Despite all the hoopla about hydrogen, Cowan will likely have been seen to have won the day when the dust settles.

For years the crux of this energy revolution hinged on these two technologies and the extraordinary possibilities they presented for a wholesale upending of the energy establishment. There were a few nagging gaps in the picture, though, but Dr. Lou Circeo, the Director of Plasma Research at Georgia Tech Research Institute, came riding in on a lightning bolt to fill them in. Lou was gracious enough to offer the expertise gained in his thirty-five years of research on plasma technologies, and the pieces he brought to the table were just what was needed to finish the puzzle.

If I have forgotten to mention some others I beg their indulgence. Countless individuals over the years have engaged me in discussions to tease out the implications of an intersection between these technologies and the politics that could bring them to fruition. My brothers Jim and Dave were especially helpful, as was my dear friend Marcus Taylor. The entire process has been immensely enjoyable. I can hardly imagine that I'll be treated to such a succession of eureka moments again, no matter how long I live. I owe you all a debt of gratitude. May you find this book worthy of your generous contributions.

~ *Tom*

Bibliography

"20% Wind Energy by 2030." US Dept. of Energy, May 2008.

(EFN), S. M. Javad Mortazavi. "High Background Radiation Areas of Ramsar, Iran." Kyoto, Japan: Kyoto University, 2002.

(IEA), International Energy Agency. "Projected Costs of Generating Electricity — 2005 Update." 124. Paris: OECD, IEA, NEA, 2005.

(PBS), Frontline. "Three Mile Island: The Judge's Ruling." In *Nuclear Reaction: Why Do Americans Fear Nuclear Power?*, Apr 22, 1997.

(UIC), Uranium Information Center. "Nuclear Power in France." Melbourne: Australian Uranium Association, April 2008.

— — —. "The Economics of Nuclear Power." Melbourne: Australian Uranium Association, Mar 2008.

Administration, Energy Information. "Residential Consumption of Electricity by End Use, 2001." DOE, 2001.

Akselsson, Mattias. "The World's Leader in Wind Power." In *Scandinavica. com*, Sep 2004.

Alex Gabbard, Oak Ridge National Laboratory. "Coal Combustion: Nuclear Resource or Danger?" Feb 5, 2008.

Allen-Mills, Tony. "Biofuel Gangs Kill for Green Profits." *The Sunday Times* Jun 3, 2007.

Allexperts. *Monju* 2007 [cited. Available from http://experts.about.com/e/m/mo/monju.htm.

Areva. "EPR Fast Facts." In *Unistar Nuclear Energy*, 2007.

Association, World Nuclear. "Advanced Nuclear Power Reactors." May 2008.

Bader, R.M. Flores and L.R. "A Summary of Tertiary Coal Resources of the Raton Basin, Colorado and New Mexico." In *U.S. Geological Survey Professional Paper 1625-A*: U.S. Geological Survey, 1999.

Barnard, Jeff. "Researchers Track Dust, Soot from China." *Boston Globe* Jul 13, 2007.

Barringer, Felicity. "California, Taking Big Gamble, Tries to Curb Greenhouse Gases." *New York Times* Sep 15, 2006.

BBC. "Asia's Greenhouse Gas 'to Treble'." In *BBC News International Edition*, Dec 14, 2006.

———. "China 'World's Worst Polluter'." BBC One-Minute World News, Apr 14, 2008.

———. "China Mulls Energy Reserves Spend." In *BBC News International Edition*, Dec 27, 2006.

———. "Climate Change 'Affecting' China." In *BBC News International Edition*, Feb 6, 2007.

Bellona.org. "Beloyarsk Nuclear Power Plant." Dec 28, 1999.

"Biofuel's Dirty Little Secret." *New Scientist* Apr 21, 2007, 7.

Biofuelwatch. "This Is Not Clean Energy: The True Cost of Our Biofuels." 2007.

Birger, John. "Mccain's Farm Flip." *Fortune* Oct 31, 2006.

Bivens, Matt. "Two-Bullet Roulette." *The Nation* Sep 10, 2003.

Bluejay, Michael. *Saving Electricity* 2007 [cited. Available from http://tinyurl.com/2nks3f.

Boyd, Robert S. "Glaciers Melting Worldwide, Study Finds." *Contra Costa Times*, Aug 21, 2002.

Bradley, Matt. "Spent Nuclear Fuel Edges Closer to Yucca." *The Christian Science Monitor* Jul 27, 2006.

Bradsher, Keith. "Emissions by China Accelerate Rapidly." *International Herald Tribune* Nov 7, 2006.

Braun, Harry W. *The Phoenix Project: Shifting from Oil to Hydrogen*: Sustainable Partners Inc, Dec 1, 2000.

Brockett, Fridley, Lin & Lin. "A Tale of Five Cities: The China Residential Energy Consumption Survey." In *Human and Social Dimensions of Energy Use: Understanding Markets and Demand*. Berkeley, CA: Lawrence Berkeley National Laboratory, 2003.

Brown, David L. *What the IPCC Report Didn't Tell Us* 2007 [cited 2007]. Available from http://starphoenixbase.com/?p=353.

Butler, Rhett A. "Soybeans May Worsen Drought in the Amazon Rainforest." In *Mongabay.com*, Apr 18, 2007.

"California Says "Go" To Energy-Saving Traffic Lights." US DOE, May 2004.

Center, International Nuclear Safety. "Overview of Fast Reactors in Russia and the Former Soviet Union." In *Internal Document*. Argonne, IL: Argonne National Laboratory, 1995.

Centre, Kansai Occupational Safety & Health. "Tokaimura Nuclear Accident from an Occupational Safety and Health Viewpoint." Japan Occupational Safety and Health Resource Centre, Apr-Jun 2001.

Chandler, David L. "Record Ice Core Reveals Earth's Ancient Atmosphere." *New Scientist*, Nov 24, 2005.

"Chernobyl Accident: Nuclear Issues Briefing Paper 22." Uranium Information Center, Feb 2008.

CHP, Mike Stabin Ph.D. "Nuclear and Radiation Safety Issues." *L.A. Times* 2006.

Circeo, Louis. *Plasma Processing of Msw at Fossil Fuel Power Plants*. Atlanta, GA: Georgia Tech Research Institute, 2007. Poster.

Circeo, Louis, and Kevin Caravati. "Plasma Processing of Msw at Fossil Fuel Power Plants." Paper presented at the HTPP9 Symposium, Orlando, FL, Feb 22, 2007.

Citro, Reuel Shinnar and Francesco. "A Road Map to U.S. Decarbonization." *Science* 313 (Sept 1, 2006): 1243.

"Climate Change Post-2100: What Are the Implications of Continued Greenhouse Gas Buildup?" In *Environmental and Energy Study Institute (EESI)*. Washington D.C., Sep 21, 2004.

"Coalbed Methane--an Untapped Energy Resource and an Environmental Concern." edited by U.S. Geological Survey, Jan 17, 1997.

Cohen, Bernard L. "The Nuclear Energy Option." edited by University of Pittsburgh: Plenum Press, 1990.

Comby, Bruno P. "Environmentalists For Nuclear Energy", TNR Editions (available on www.ecolo.org), 2001.

Commission, European. *Research Team of the Externe Project Series* 2007 [cited. Available from http://www.externe.info/team.html.

Commission, N.Z. Electricity. "Electricity Efficiency Can Influence Future Load Growth." Sep 5, 2005.

Communications, Nebraska College of Journalism & Mass. "Most of Nebraska Corn Crop Will Go to Ethanol by 2011." In *News Net Nebraska*, Aug 30, 2006.

"Congressman Harris Fawell (R-IL)." Congressional Record: August 10, 1994.

Constantino, Renato Redentor. "With Nature There Are No Special Effects." In *TomDispatch.com*, June 3, 2004.

Consultancy, Solarbuzz Research &. *Solar Photovoltaic Electricity Price Index* 2007 [cited. Available from http://www.solarbuzz.com/SolarIndices.htm.

Corporation, CNA. "National Security & the Threat of Climate Change." Alexandria, VA, 2007.

Coughlin, Kevin. "King Coal Comes Clean." *The Star-Ledger* Mar 6, 2005.

Cowan, Graham. "Boron: A Better Energy Carrier Than Hydrogen?" In *11th CHC Hydrogen Research Conference*, 2002.

— — —. *Boron: A Better Energy Carrier Than Hydrogen?* 2007 [cited. Available from http://www.eagle.ca/~gcowan/boron_blast.html.

Davis, Christopher M. "9/11 Commission Recommendations: Joint Committee on Atomic Energy — a Model for Congressional Oversight?" edited by Government and Finance Division, Aug 20, 2004.

Donald E. Lutz, P. E. "PG&E Solar Plants in the Desert." In *Truth About Energy*, 2007.

— — —. "Wind." In *Truth About Energy*, 2007.

Donovan, Patrick H. "Oil Logistics: In the Pacific War." *Air Force Journal of Logistics* (Spring 2004).

Dubberley, Allen E. "S-Prism Fuel Cycle Study." Paper presented at the International Congress on Advances in Nuclear Power Plants (ICAPP), Cordoba, Spain May 4-7, 2003.

Econexus. "Petition Calling for an Agrofuel Moratorium in the E.U." 2006.

Editorial. "We Cannot Afford This Extra Pollution." *The Daily Astorian* Jun 13, 2006.

Elobeid, Tokgoz, Hayes, Babcock, & Hart. "The Long-Run Impact of Corn-Based Ethanol on the Grain, Oilseed, and Livestock Sectors: A Preliminary Assessment." Center for Agricultural and Rural Development, Iowa State University, Nov 2006.

EPA, U.S. "Sources of GHGs in the U.S." Pew Center on Global Climate Change, 1998.

Exec Director, Eric S. Beckjord. "The Future of Nuclear Power." Massachusetts Institute of Technology, 2003.

Fairley, Peter. "Part I: China's Coal Future." *MIT Technology Review* (Jan 4, 2007).

Feature. "Japan's Safety Response." *Nuclear Engineering International* Dec 20, 2006.

Forum, US DOE Nuclear Energy Research Advisory Committee & Gen IV International. "A Technology Roadmap for Generation Iv Nuclear Energy Systems." 40: US Dept. of Energy, Dec 2002.

fueleconomy.gov. *Advanced Technologies & Energy Efficiency* US DOE/EPA, 2007 [cited. Available from http://www.fueleconomy.gov/FEG/atv.shtml.

Gavin Schmidt, Michael Mann. "Decrease in Atlantic Circulation?" In *Real Climate*, Nov 30, 2005.

Gellerman, Bruce. "Don't Mess with Texas Wind." In *Living on Earth*. USA, Jun 6, 2008.

Gongloff, Mark. "Keeping up with T. Boone." *The Wall Street Journal* Sep 19, 2007.

Goodell, Jeff. "Can Dr. Evil Save the World?" *Rolling Stone* Nov 3, 2006.

Greenpeace. "Chernobyl Death Toll Grossly Underestimated." Apr 18, 2006.

Hanley, Charles J. "U.N. Nations Reach Deal to Cut Emissions." *Washington Post* Nov 17, 2006.

Harvey, Fiona. "Beware the Carbon Offsetting Cowboys." *Financial Times* Apr 25, 2007.

Hirsch, Tom. "Brazilian Biofuels' Pulling Power." In *BBC News International Edition*, Mar 8, 2007.

Hirshon, Bob. "Asian Brown Cloud." In *Science Netlinks (AAAS)*, Jan 12, 2003.

Holdren, John P. *Global Energy Challenges & the Role of Increased Energy Efficiency in Addressing Them*. Berkeley, California, 2006.

Holson, Richard A. Oppel & Laura M. "While a Utility May Be Failing, Its Owner Is Not." *New York Times* Apr 30, 2001.

Hooper, Rowan. "Uk Wind Power Takes a Battering." *New Scientist* Nov 12, 2005.

Hoopman, Dave. "Home-Grown Energy Sources Continue to Expand." In *Wisconsin Energy Cooperative News*, Jan 2007.

IAEA. "Nuclear Desalination." In *Global Development of Advanced Nuclear Power Plants*, 14. Vienna: International Atomic Energy Agency, Sep 2006.

— — —. "Operational & Decommissioning Experience with Fast Reactors." Cadarache, France, Mar 11-15, 2002.

ICAPP. "Proceedings of ICAPP '03." Paper presented at the International Congress on Advanced Nuclear Power Plants, Cordoba, Spain May 4-7, 2003.

IEA. "30 Key Energy Trends in the IEA & Worldwide." 33. Paris: International Energy Agency, 2001.

— — —. *China's Power Sector Reforms*. Paris: International Energy Agency, 2006.

— — —. "Costs for Different Renewables." BBC News, 2004.

— — —. "Renewables in Global Energy Supply, an IEA Fact Sheet." International Energy Agency, Jan 2007.

— — —. "World Energy Outlook 2004." 62. Paris: International Energy Agency, 2004.

Incineration, Committee on Health Effects of Waste. "Waste-Incineration and Public Health." edited by National Research Council, 3: National Academy of Sciences, 2000.

"Incinerators in Disguise." Global Alliance for Incinerator Alternatives, April 2006.

Institute, Worldwatch. "Phasing out Coal: Environmental Concerns, Subsidy Cuts Fuel Global Decline." Aug 25, 1999.

"Inventory of U.S. Greenhouse Gas Emissions and Sinks: 1990–2004." US EPA, Apr 15,2006.

IPCC. "Underground Geological Storage." In *IPCC Special Report on Carbon Capture & Storage*, Chapter 5, Sep 2005.

IPCC, Working Group 1 of the. "Climate Change 2007: The Physical Science Basis." Geneva, Switzerland: Intergovernmental Panel on Climate Change (IPCC), 2007.

James. *Clean Coal or Dirty Coal?* Oct 3, 2006 [cited. Available from http://tinyurl.com/3wxh39.

Johnsrud, Judith. "Why Nuclear Power Is Not a Solution." *The Sylvanian* Nov 2005-Jan 2006.

Kachan, Dallas. "$13b Nuclear Fusion Research Agreement Signed." In *Cleantech.com*, Nov 21, 2006.

Ken Zweibel, James Mason and Vasilis Fthenakis. "A Solar Grand Plan." *Scientific American* January 2008.

Kleiner, Kurt. "Powdered Metal: The Fuel of the Future." *New Scientist* Oct 22, 2005.

Komanoff, Charles. "Don't Trade Carbon, Tax It." In *Grist Environmental News & Commentary*, Feb 13, 2007.

Kosanovic, Lisa. "Clean Coal? New Technologies Reduce Emissions, but Sharp Criticisms Persist." *E: The Environmental Magazine* Jan-Feb 2002.

Laboratory, Idaho National. "The Future of Geothermal Energy." 1-21: Renewable Energy & Power Dept. of INL, 2006.

Lazaroff, Cat. "Coal Burning Power Plants Spewing Mercury." In *Environment New Service*, Nov 18, 1999.

Lederer, Edith M. "Un Expert Seeks to Halt Biofuel Output." *Associated Press*, October 26, 2007.

Lenntech, (Delft Technical University). "The Global Warming and the Greenhouse Effect." 2007.

Lepisto, Christine. "Geothermal Power Plant Triggers Earthquake in Switzerland." In *Treehugger.com*, Jan 21, 2007.

Lewis, Judith. "Green to the Core? — Part 1." In *L.A. Weekly*, Nov 10, 2005.

Lipman, Zada. "A Dirty Dilemma." *Harvard International Review* 23, no. 4 (Winter 2002).

Little, Amanda Griscom. "Coal Position." *Grist Environmental News & Commentary* (Dec 3, 2004).

Lochbaum, David. "Regulatory Malpractice: The Nrc's Handling of the Pwr Containment Sump Problem." Union of Concerned Scientists, 2003.

Lundegaard, Jeffrey Ball & Karen. "Loophole Gives SUV Buyers a Tax Break." *Salt Lake Tribune* Dec 20, 2002.

Martin, L.J. "Carbon Neutral - What Does It Mean?" In *Ezine Articles*, Oct 26, 2006.

McErlain, Eric. "Real Science Refutes The "Tooth Fairy"." In *NEI Nuclear Notes*, Mar 4, 2005.

McGrail, D. H. Bacon & B. P. "Waste Form Release Calculations for the 2005 Integrated Disposal Facility Performance Assessment." Pacific Northwest National Laboratory, July 2005.

Meyerson, Harold. "Power to (and from) the People!" *L.A. Weekly* Feb 2, 2001.

Mintz, Folga, Molburg, Gillette. "Cost of Some Hydrogen Fuel Infrastructure Options." Transportation Technology R&D Center, Argonne National Laboratory, Jan 16, 2002.

Moore, John Dunbar & Robert. "California Utilities' Donations Shed Light on Blackout Crisis." In *Center for Public Integrity*, May 30, 2001.

Munson, Richard. "Yes, in My Backyard: Distributed Electric Power." In *Issues in Science & Technology*, Winter 2006.

Netherlands, Radio. "Brown Pall over Asia." Aug 12, 2002.

News, CBS Evening. "The Most Polluted Places on Earth." June 6, 2007.

O'Leary, Wayne. "Electricity Illuminates the Ghost of George Norris." *The Progressive Populist* Jan 2006.

Olson, Thomas. "Gamble on Plasma Turns into Jackpot." *Pittsburgh Tribune-Review* Apr 26, 2008.

Painter, James. "Peru's Alarming Water Truth." In *BBC News International Edition*, Mar 12, 2007.

Palast, Greg. "Blood in Beirut: $75.05 a Barrel." In *The Guardian*, 2006.

— — —. *It's Still the Oil* 3/18/2007 [cited. Available from http://tinyurl. com/2vuesr.

Palast, Gregory. "Some Power Trip." *Washington Post* Jan 28, 2001.

Paul, Jim. "Experts: Ethanol's Water Demands a Concern." In *MSNBC*, Jun 18, 2006.

Pearce, Fred. "Climate Warning as Siberia Melts." *New Scientist*, Aug 11, 2005, 12.

— — —. "Fuels Gold: Big Risks of the Biofuel Revolution." *New Scientist* Sep 25, 2006, 36-41.

Pegg, J.R. "Coal Power Soot Kills 24,000 Americans Annually." In *Environment New Service*, Jun 10, 2004.

Ph.D., Kirk Smith. *Health Burden from Indoor Air Pollution in China*. Berkeley, CA: World Health Organization, 2003.

Port, Otis. "Power from the Sunbaked Desert." *Business Week* Sep 12, 2005.

Press, Associated. "France and 45 Other Countries Call for World Environmental Monitor." *International Herald Tribune* Feb 4, 2007.

"Profile: Hugo Chavez." In *BBC News International Edition*, Dec 3, 2007.

Program, UN Development. "Denmark: The Human Development Index - Going Beyond Income." UNDP Human Development Reports, Nov 27, 2007.

Public_Citizen. *New Nuclear Power Plants = More Nuclear Waste* Aug 2003 [cited. Available from http://tinyurl.com/5lps7a.

Quarterly, Technology. "Tapping the Oceans." *The Economist* Jun 5, 2008.

R.F. Blundy, P.G. Zionkowski. "Final Report, "Demonstration of Plasma in Situ Vitrification at the 904-65g K-Reactor Seepage Basin."", edited by DOE. Aiken, SC: Westinghouse Savannah River Company, Dec 1997.

Ramey, David E. *Butyl Fuel, Llc* [cited 2007]. Available from http://www. butanol.com/index.html.

Reef Ball Foundation [cited 2007]. Available from http://www.reefball.org/.

Release, Argonne National Laboratory News. "Argonne Fast-Reactor Pioneer Receives International Prize." May 7, 2004.

Reuters. "BC Transmission Corporation Chooses 3M's Aluminum Matrix Conductor for Two Segments of Vancouver Island Transmission Reinforcement Project." May 14, 2008.

— — —. "New PQ Media Report: Campaign Media Spending to Grow 64.1% in 2008 to All-Time High." Dec 6, 2007.

Ridgeway, Daniel Schulman & James. "The Highwaymen." *Mother Jones* Jan 1, 2007.

Robert S. Norris, William M. Arkin. "World Plutonium Inventories - 1999." *Bulletin of the Atomic Scientists* Sept-Oct 1999.

Roberts, Paul. "The Undeclared Oil War." *Washington Post*, June 28, 2004, A21.

Romm, Salon.com Interview w/Joseph. "Just Say No, to Hydrogen." In *Salon.com*, Apr 29, 2004.

Rosenberg, Larry, and Christine Furedy. "International Source Book on Environmentally Sound Technologies for Municipal Solid Waste Management." 1.5.4: UNEP Environmental Technology Center, 1996.

— — —. "International Source Book on Environmentally Sound Technologies for Municipal Solid Waste Management." 1.5.1: UNEP Environmental Technology Center, 1996.

Sample, Ian. "Scientists Offered Cash to Dispute Climate Study." *The Guardian*, Feb 2, 2007.

— — —. "Warming Hits 'Tipping Point'." *The Guardian*, Aug 11, 2005.

SEISCO. "Comparison of Estimated Annual Water Heating Costs." 2007.

"Senator Paul Simon (D-IL)." Congressional Record-Senate: June 30, 1994.

Service, Nuclear Information & Resource. "Think Nuclear Power Can Save the Climate?" NIRS, Oct 27, 2006.

Shah, Sonia. "The End of Oil? Guess Again." In *Salon.com*, Sep 15, 2004.

Sheet, Fact. "Final Air Regulations for Municipal Solid Waste Landfills." edited by US EPA, Mar 1, 1996.

Sladky, Lynne. "Florida County Plans to Vaporize Landfill Trash." *USA Today* 9/9/2006.

Slocum, Tyson. "Big Oil Can Afford to Forgo Tax Breaks — but Renewable Energy Can't." In *Public Citizen*, Feb 27, 2008.

Smith, Kevin. "'Obscenity' of Carbon Trading." In *BBC News International Edition*, Nov 9, 2006.

"Spent Nuclear Fuel Returned to the United States from Germany." edited by U.S. Dept of Energy: National Nuclear Security Administration, Sept 2004.

Stanford, George S. "Integral Fast Reactors: Source of Safe, Abundant, Non-Polluting Power." In *National Policy Analysis*, Dec 2001.

Statemaster.com. "Total Electricity Consumption (Per Capita) by State." National Priorities Project Database, 2001.

Steenblik, Richard Doornbosch and Ronald. "Biofuels: Is the Cure Worse Than the Disease?" In *Round Table on Sustainable Development*: OECD, Sep 11-12, 2007.

Steger, Will. *Global Warming 101.Com* Will Steger Foundation, 2006 [cited 2007]. Available from http://www.globalwarming101.com/content/view/545/88889028/.

Steve Connor, Jonathan Brown. "Tackle Nuclear Waste Disposal First, Warn Advisers." *The Independent* Jan 24, 2006.

Strahan, David. "The Great Coal Hole." *New Scientist*, Jan 19, 2008, 38-41.

Strickland, Jonathan. *How Plasma Converters Work* 2007 [cited May 3 2007]. Available from http://science.howstuffworks.com/plasma-converter2.htm.

Subcommittee, APS Panel on Public Affairs Energy. "Hydrogen Initiative Report from American Physical Society Panel Released." American Physical Society, Mar 1, 2004.

Survey, US Geological. *Volcanic Lakes and Gas Releases* 1999 [cited. Available from http://tinyurl.com/4fd7fa.

"Technology Options for the near and Long Term ", 34: U.S. Climate Change Technology Program, Aug 2005.

Television, Kentucky Educational. *KET Coal Mine Field Trip Q & A* Dec 15, 2005 [cited. Available from http://www.ket.org/Trips/coal/qa.html.

"Testimony of Kelly Fletcher of GE." In *U.S. Senate Energy & Water Subcommittee.* Washington, DC: General Electric, Sep 14, 2006.

The End Is Sigh Grist Environmental News & Commentary, Nov 20, 2006 [cited. Available from http://www.grist.org/news/daily/2006/11/20/.

Till, Dr. Charles. "Plentiful Energy, the IFR Story, and Related Matters." *The Republic* Jun-Sep 2005.

"Total Midyear Population for the World: 1950-2050." U.S. Census Bureau, 2007.

Varian, Hal R. "Recalculating the Costs of Global Climate Change." *New York Times* Dec 14, 2006.

Wikipedia. "Little Ice Age." 2008.

Wilson, Jeff. "Corn Rises as Argentina Halts Exports to Conserve Supplies." In *Bloomberg.com*, Nov 20, 2006.

WNA. "Fast Neutron Reactors." In *Information Papers.* London: World Nuclear Association, Feb 2008.

—— . "Nuclear Power in China." World Nuclear Association, Apr 2008.

Glossary

€ents: One one-hundredth of a euro; a eurocent. Coined for our purposes in this book, since there doesn't seem to be a ¢ symbol equivalent for the euro.

actinide: The 14 chemical elements that lie between actinium and nobelium (inclusively) on the periodic table, with atomic numbers 89–102. Only actinium, thorium, and uranium occur naturally in the earth's crust in anything more than trace quantities. Plutonium and others are man-made actinides resulting from neutron capture, produced mainly in fission reactions.

ABWR: Advanced Boiling Water Reactor. A Generation III lightwater reactor built by GE/Hitachi, with substantial safety and economy improvements compared to Gen II reactors.

AHTR: Advanced High Temperature Reactor. An experimental reactor being designed to operate at temperatures high enough to produce hydrogen via the thermo-chemical sulfur-iodine cycle. AHTRs may exacerbate the nuclear waste problem, however, by producing spent fuel that is difficult or impossible to recycle.

ALMR: Advanced Liquid Metal Reactor. The type of fast reactor used in an IFR power plant. Molten sodium is the liquid-metal coolant. See "LMR," below.

AP-1000: A Generation III lightwater reactor from Westing-house that utilizes modular construction and passive safety systems similar to those that will be employed in IFRs.

AREVA: France's nuclear power agency that oversees all aspects of the process, from mining to waste disposal.

B/G: biogarbage. Used to denote fuels derived from garbage and other waste products via plasma converters

beta decay: In beta decay, a neutron is converted into a proton while emitting an electron and an anti-neutrino. Because the number of protons in the nucleus is different for each element, beta decay actually changes one element into another.

BWR: Boiling Water Reactor: a type of light water nuclear reactor developed in the 1950s by General Electric

CAFE standards: Corporate Average Fuel Economy, a U.S. standard for minimum fuel efficiency in vehicles

CANDU: CANada Deuterium Uranium. A type of nuclear power reactor using heavy water as a moderator and natural, unenriched uranium as fuel.

CCL: Cold cathode lighting

CCT: clean coal technologies

CFL: Compact fluorescent light bulbs

DOE: U.S. Department of Energy.

EBR-II: The Experimental Breeder Reactor that demonstrated the feasibility and safety of the IFR concept, successor to the earlier EBR-I, of course.

EFN: Environmentalists For Nuclear Energy – www.ecolo.org

EPA: the U.S. Environmental Protection Agency

EPR: European Pressurized Reactor. The most up-to-date design of light water reactor, a so-called Third Generation plant incorporating the latest in safety features and modern technology. The USA version uses the same acronym but considers it to signify Evolutionary Pressurized Reactor. Silly, no?

ESBWR: Economic Simplified Boiling Water Reactor. GE's answer to the AP-1000, a transitional LWR employing passive safety principles and modular construction.

fast reactor: A nuclear reactor in which the neutrons are not moderated, instead remaining energetic. Fast reactors can consume all actinides completely; thermal reactors cannot.

GHG: greenhouse gases. The primary culprit in terms of manmade emissions is carbon dioxide, but other important ones are methane and oxides of nitrogen.

GNEP: Global Nuclear Energy Partnership. A U.S. proposal for an international agreement to develop proliferation-resistant power systems, including fast reactors, around the world. GNEP seeks to develop a worldwide consensus on enabling expanded use of economical, carbon-free nuclear energy to responsibly meet the world's burgeoning energy demands.

GREAT: Global Rescue Energy Alliance Trust. This author's proposed international organization to oversee all aspects of nuclear power worldwide. Very similar to France's AREVA, only expanded to an international level.

GW: gigawatt, equal to a thousand megawatts. A typical large power plant would produce about one gigawatt of electric power. See "kWe."

half-life: the amount of time required for a radioactive substance to decay to half its quantity. As radioactive elements decay toward a stable state their radioactivity decreases, eventually falling below the harmless levels of normal background radiation.

HWR: heavy water reactor. A nuclear reactor that uses deuterium oxide (aka heavy water) as its coolant and as a moderator to reduce the velocity of fast neutrons, as in CANDU reactors. See "thermal reactor," below.

IAEA: International Atomic Energy Agency

IEA: International Energy Agency: an organization made up of about 150 energy experts and statisticians from 26 member countries, who since 1974 have acted in an advisory capacity on energy issues for countries both in and outside that group.

IFR: Integral Fast Reactor. A fast-reactor plant that incorporates a pyroprocessing system on-site for closing the fuel cycle, assuring that weapons-grade material will never be separated out and that actinides will never leave the premises unless needed as startup fuel for new fast reactors.

IGCC: Integrated Gasification Combined Cycle, a "clean coal" technology

IPCC: Intergovernmental Panel on Climate Change. A U.N. body established in 1988 to evaluate the risk of climate change brought on by humans, based mainly on peer reviewed and published scientific/technical literature.

IRIS: International Reactor Innovative & Secure. A transitional modular reactor to bridge the gap between Gen III and Gen IV reactors, this is being designed by an international consortium with certification expected by about 2015.

ISPV: In-Situ Plasma Vitrification. The process of boring holes into the earth to break down compounds underground and produce a stable vitreous layer using subterranean plasma torches.

ITER: International Thermonuclear Experimental Reactor, a fusion research program that has been on-again/off-again since 1985.

JCAE: the U.S. Congress' Joint Committee on Atomic Energy, 1946–1977

JCEC: Joint Committee on Energy and Climate, the author's suggested joint supercommittee patterned on the JCAE

kWe: kilowatt (electric). Power generation is sometimes designated in watts with a "t" or "e" after the unit, indicating thermal energy or electrical energy. In a power plant that uses heat to generate electricity, the "t" number will always be higher than the electricity "e" it produces, due to the energy penalty of converting from heat to electricity.

LMR: Liquid Metal Reactor: A fast reactor cooled by a liquid metal, such as sodium, lead, or a lead-bismuth alloy. See "ALMR," above

LWR: Light Water Reactor: a nuclear reactor using regular water (as opposed to "heavy water") as a moderator to slow down the neutrons during the fission process. Most reactors in use today are LWRs.

moderator: a substance used to slow down ("moderate") the neutrons in a thermal reactor. The most commonly used moderators are light water, heavy water, and graphite.

MSW: municipal solid waste, aka garbage

MW: megawatt, equal to one million watts or a thousand kilowatts. See "kWe"

Newclear power: Integral Fast Reactor (IFR) technology, termed "*new*clear" to emphasize its vast improvements over "*old*" presently deployed nuclear power technology such as LWRs, heavy water, and graphite-moderated reactors.

NRC: U.S. Nuclear Regulatory Commission, tasked with design certification and oversight of all civilian nuclear power plants in the USA.

OECD: Organization for Economic Cooperation & Development: a group of 30 member countries who discuss and develop economic and social policy. These countries account for two-thirds of the world's goods and services.

OPEC: Organization of Petroleum Exporting Countries. The international oil cartel.

ORNL: Oak Ridge National Laboratory, in Tennessee

PG&E: Pacific Gas & Electric, one of the two main utility companies in California, the other being SCE.

PRISM: Power Reactor Innovative Small Module. An advance liquid metal reactor designed by General Electric, the type of reactor that would be coupled with pyroprocessing facilities that together would make up an IFR.

Pu: the chemical symbol for plutonium

PV: Photovoltaics (solar cells)

PWR: Pressurized Water Reactor, a type of LWR.

pyroprocessing: a non-aqueous industrial process utilizing high temperatures to effect chemical and physical changes. This book describes how pyroprocessing can be used for the on-site recycling of the spent fuel from fast reactors to recover the remaining actinides and incorporate them into the new fuel assemblies for reinsertion into the reactor.

S-PRISM: Super-PRISM. The scaled-up version of GE's PRISM reactor.

SCE: Southern California Edison, one of the two main utility companies in California, the other being PG&E.

thermal reactor: a nuclear reactor that uses either ordinary "light" water, heavy water, or graphite to slow the neutrons emitted from its fuel in order to increase their odds of fissioning with another atom of fuel. Nearly every nuclear power plant in use today is a thermal reactor of one kind or another: LWR, BWR, EPR, PWR, CANDU, etc.

TW: terawatt, equal to a thousand gigawatts (GW)

U: the chemical symbol for uranium

UNSCEAR: United Nations Scientific Commission on the Effects of Atomic Radiation, the UN body with a mandate from the General Assembly to assess and report levels and health effects of exposure to ionizing radiation.

WHO: World Health Organization, an agency of the United Nations, established in 1948, concerned with improving the health of the world's people and preventing or controlling communicable diseases on a worldwide basis through various technical projects and programs.

WRI: World Resources Institute

WWF: World Wildlife Fund

Made in the USA
Lexington, KY
16 November 2011